Advances in Intelligent and Soft Computing

153

For further volumes:
http://www.springer.com/series/4240

Paulo Novais, Kasper Hallenborg, Dante I. Tapia,
and Juan M. Corchado Rodríguez (Eds.)

Ambient Intelligence - Software and Applications

3rd International Symposium on Ambient Intelligence (ISAmI 2012)

Editors
Paulo Novais
Universidade do Minho
Departamento de Informática
Campus de Gualtar
Braga
Portugal

Dante I. Tapia
Departamento de Informática y Automática
Facultad de Ciencias
Universidad de Salamanca
Salamanca
Spain

Kasper Hallenborg
The Maersk Mc-Kinney Moller Institute
University of Southern Denmark
Odense
Denmark

Juan M. Corchado Rodríguez
Departamento de Informática y Automática
Facultad de Ciencias
Universidad de Salamanca
Salamanca
Spain

ISSN 1867-5662 e-ISSN 1867-5670
ISBN 978-3-642-28782-4 e-ISBN 978-3-642-28783-1
DOI 10.1007/978-3-642-28783-1
Springer Heidelberg New York Dordrecht London

Library of Congress Control Number: 2012933126

Printed on acid-free paper

Springer is part of Springer Science+Business Media (www.springer.com)

Preface

This volume contains the proceedings of the 3th International Symposium on Ambient Intelligence (ISAmI 2012). The Symposium was held in Salamanca, Spain on March 28th–30th at the University of Salamanca, under the auspices of the Bioinformatic, Intelligent System and Educational Technology Research Group (http://bisite.usal.es/) of the University of Salamanca.

The ISAmI conference has been running annually and provides an international forum to present and discuss the latest results, innovative projects, new ideas and research directions, and to review current trends on Ambient Intelligence namely in terms of software and applications, and aims to bring together researchers from various disciplines that are interested in all aspects of this area.

Ambient Intelligence is a recent paradigm emerging from Artificial Intelligence, where computers are used as proactive tools assisting people with their day-to-day activities, making everyone's life more comfortable.

ISAmI 2012 received a total submission of 48 papers from 17 different countries. Each paper has been reviewed by, at least, three different reviewers, from an international committee composed of 51 members from 20 countries. After a careful review, 22 papers were selected as long papers and 5 were selected as short papers to be presented at the conference and published in the proceedings.

A Special Session on Artificial Intelligence Technologies in Resource-Constrained Devices was organized to complement the regular program, 4 long papers were accepted in the session.

A Thematic Issue in the Journal of Ambient Intelligence and Smart Environments (IOS Press) will be edited.

A special thanks to Ad van Berlo from the Dutch national expert centre for Smart Houses and Smart Living that was our invited speaker with the talk "From smart home technology towards ambient assisted living - lessons learnt".

We want to thank all the sponsors of ISAmI'12: IEEE Sección España, CNRS, AFIA, AEPIA, APPIA, and Junta de Castilla y León.

We would like also to thank all the contributing authors, as well as the members of the Program Committee and the Local Organizing Committee for their hard and highly valuable work.

Your work was essential to the success of ISAmI 2012.

March 2012

The Editors
Paulo Novais
Kasper Hallenborg
Dante I. Tapia
Juan M. Corchado Rodríguez

Organization

Scientific Committee Chairs

Paulo Novais	University of Minho, Portugal
Kasper Hallenborg	University of Southern Denmark, Denmark

Organizing Committee Chair

Juan M. Corchado	University of Salamanca, Spain
Dante I. Tapia	University of Salamanca, Spain

Steering Committee

Juan Carlos Augusto	University of Ulster, UK
Juan M. Corchado	University of Salamanca, Spain
Paulo Novais	University of Minho, Portugal

Program Committee

Andreas Riener	Johannes Kepler University Linz, Austria
Antonio F. Gómez Skarmeta	University of Murcia, Spain
Antonio Fernández Caballero	University of Castilla-La Mancha, Spain
Benjamín Fonseca	Universidade de Trás-os-Montes e Alto Douro, Portugal
Carlos Bento	University of Coimbra, Portugal
Carlos Juiz	University of the Balearic Islands, Spain
Carlos Ramos	Polytechnic of Porto, Portugal
Cecilio Angulo	Polytechnic University of Catalonia, Spain
Cristina Buiza	Ingema, Spain
Davy Preuveneers	Katholieke Universiteit Leuven, Belgium
Diane Cook	Washington State University, USA
Eduardo Dias	New University of Lisbon, Portugal
Elisabeth Eichhorn	Carmeq GmbH, Germany
Emilio S. Corchado	University of Burgos, Spain
Flavia Delicato	Universidade Federal do Rio de Janeiro, Brasil
Florentino Fdez-Riverola	University of Vigo, Spain
Fluvio Corno	Politecnico di Torino, Italy
Francesco Potortì	ISTI-CNR Institute, Italy

Francisco C. Pereira	SMART (Singapore-MIT Alliance), Singapore
Francisco Silva	Maranhão Federal University, Brazil
Goreti Marreiros	Polytechnic of Porto, Portugal
Gregor Broll	DOCOMO Euro-Labs, Germany
Guillaume Lopez	University of Tokyo, Japan
Habib Fardoum	University of Castilla-La Mancha, Spain
Hans W. Guesgen	Massey University, New Zealand
Ichiro Satoh	National Institute of Informatics Tokyo, Japan
Jaderick Pabico	University of the Philippines Los Baños, Philippines
Javier Jaen	Polytechnic University of Valencia, Spain
Jo Vermeulen	Hasselt University, Belgium
José M. Molina	University Carlos III of Madrid, Spain
José Machado	University of Minho, Portugal
José Maria Sierra	Universidad Carlos III de Madrid, Spain
Joyca Lacroix	Philips Research, Netherlands
Junzhong Gu	East China Normal University, China
Kevin Curran	University of Ulster, UK
Kristof Van Laerhoven	TU Darmstadt, Germany
Lourdes Borrajo	University of Vigo, Spain
Martijn Vastenburg	Delft University of Technology, Netherlands
Pablo Haya	Universidad Autónoma de Madrid, Spain
Radu-Daniel Vatavu	University "Stefan cel Mare" of Suceava, Romania
Rene Meier	Trinity College Dublin, Ireland
Ricardo Costa	Polytechnic of Porto, Portugal
Ricardo S. Alonso	University of Salamanca, Spain
Rui José	University of Minho, Portugal
Simon Egerton	Monash University, Malaysia
Teresa Romão	New University of Lisbon, Portugal
Tibor Bosse	Vrije Universiteit Amsterdam, Netherlands
Veikko Ikonen	VTT Technical Research Centre, Finland
Vic Callahan	University of Essex, UK
Wolfgang Reitberger	Vienna University of Technology, Austria
Yi Fang	Purdue University, USA

Additional Reviewers

Ângelo Costa
Davide Carneiro
Filipe Rodrigues
Gerold Hoelzl
Joao Oliveirinha
Marc Kurz
Marilia Curado
Raphael Machado

Local Organization Committee

Juan M. Corchado	University of Salamanca, Spain
Dante I. Tapia	University of Salamanca, Spain
Javier Bajo	Pontifical University of Salamanca, Spain
Fernando de la Prieta	University of Salamanca, Spain
Cesar Analide	University of Minho, Portugal

Special Session on Artificial Intelligence Technologies in Resource-Constrained Devices

Organizing Committee Chair

Juan Botía Blaya	University of Murcia, Spain
Joaquin Canada-Bago	Universidad de Jaén, Spain
Jose Angel Fernandez-Prieto	University of Jaén, Spain
Manuel Angel Gadeo-Martos	University of Jaén, Spain
Alicia Triviño-Cabrera	University of Malaga, Spain
Juan Ramón Velasco	University of Alcalá, Spain

Program Committee

Fernando Boavida	University of Coimbra, Portugal
Fábio Buiati	Complutense University of Madrid, Spain
Mario Collota	Universita degli Studi di Enna "KORE", Italy
Marilia Curado	University of Coimbra, Portugal
Isabelle Demeure	Telecom Paris Tech, France
James P. Dooley	Essex University, United Kingdom
Silvana Greco Polito	University of Enna "KORE", Italy
Jorge J. Gomez Sanz	Complutense University of Madrid, Spain
Gordon Hunter	Kingston University, UK
Diego López de Ipiña	University of Deusto, Spain
Luis Magdalena	European Centre for Soft Computing
Miguel Wister	University Juarez de Tabasco, México

Contents

Part I: Long Papers

Part I
Long Papers

Contacting the Devices: A Review of Communication Protocols

Álvaro San-Salvador and Álvaro Herrero

Abstract. When building an intelligent environment able to interact and assist people in a smooth and kind way, one of the practical issues to be considered is the communication technology that will connect the "brain" (intelligence) of the system with its "hands" (the different devices/actuators that the system will be equipped with). This paper reviews some of the currently available protocols, standards and technologies for such a purpose, comprehensively analyzing several key issues for comparative purposes. This comparative study supports inexperienced developers when deciding which technology best suits the requirements of the system to be deployed and what the costs of the licenses will approximately be.

Keywords: Ambient-Assisted Living, Intelligent Environment, Domotics, Communication Protocols, Protocol Features, Ambient Intelligent Devices.

1 Introduction

In order to achieve a real ambient-assisted home, intelligent devices are required, and furthermore, this kind of devices need to speak the same language and in a common channel. However, a lack of clarity has been previously identified regarding the communication aspect of such systems. It has been previously stated that today, there is a confusing set of choices and a complex hierarchy of standards from low-level specification of data formats to system level application interoperability standards [1]. In keeping with this idea, some other authors [2], [3] claimed that if a real common open standard had been developed and the compatibility of the intelligent devices had been assured, this kind of technology would be much more popular at present time.

Álvaro San-Salvador · Álvaro Herrero
Civil Engineering Department, University of Burgos. C/ Francisco de Vitoria s/n, 09006 Burgos, Spain
e-mail: asi0004@alu.ubu.es, ahcosio@ubu.es

P. Novais et al. (Eds.): Ambient Intelligence - Software and Applications, AISC 153, pp. 3–10.
springerlink.com

Nowadays, there is a wide range of standards such as X10 [4] and commercial alliances as KNX [5] or Zigbee [6] that claim to be the best possible choice to harmonize the communication of heterogeneous devices. However, some features of such systems may be restrictive, preventing its usage under certain circumstances. This study presents an objective comparison as an ongoing research to improve AmI systems' communication and ease development.

In section 3, 4 and 5 up-to-date communication solutions are described and analyzed taking into account the following key issues: type of the protocol, connectivity (Twisted Pair, Power Line Connection, Radio Frequency such as Bluetooth, Wi-Fi, etc.), economical issues, updating of the documents and security matters. A summary of the most important features under analysis can be found in section 6.

2 Previous Work

A starting-point when trying to study the communication needs and problems of an Ambient Intelligent (AmI) system is to analyze the available technologies, in the way some authors previously did [7]. Communication protocols have been previously examined and compared [8], mainly from a Quality of Service standpoint.

There are some previous papers in this direction that makes a comparative study of different wireless protocols (Bluetooth, UWB, ZigBee, and Wi-Fi) [7], that presents a theoretical performance comparison of wireless and power line networks technologies[8], or that analyzes architectures for ubiquitous systems based on the structure of the environment and the way discovery mechanisms operate [9].

Some other papers related to this topic can be found, such as: [10], which describes a service-oriented solution framework designed for home environments, and [11], which proposes a number of challenges that must be overcome, according to the authors, before the smart home concept can move to reality.

Differentiating from the above described previous work, present study focuses on the present standards trying to fulfill the requirement of a common platform to communicate with every single device in an intelligent ambient. Initially, five technologies were selected, namely X-10 [4], KNX [5] , Zigbee [6], Lon Talk [12] and Z-Wave [13]. Those were chosen as they are the most widely used at present time because they incorporate important facilities in comparison with more simple communication protocols (IP, Wi-Fi, Bluetooth, etc.).

According to the analyzed previous work, it is needed a review of the existing protocols and standards taking into account issues different from Quality of Service. This paper intends to provide with useful information that helps anyone to decide which technology best fits its personal needs in the area of AmI and Smart Homes.

3 Early Protocols: X-10

X-10 [14] was one of the very first protocols used in domotics. It was designed in 1975 by Pico Electronics (Scotland) for data transmission through home power line wiring. It is now an open standard; so every manufacturer can develop its own X-10 capable devices, but obliged to use the circuits of the company that originally developed the standard. In Fig. 1, the X-10 communication stack is depicted.

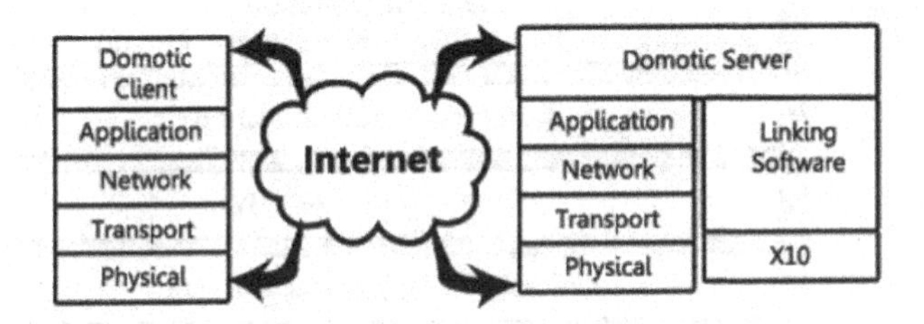

Fig. 1 X-10 protocol stack.

This protocol has some problems and limitations. It is worth mentioning the incompatibility between devices according to the region where they were bought. As an example, a device from the USA will never work properly in Europe and the other way round due to the different line frequencies. Other important limitation from the AmI standpoint is its slow response. This will cause a bad user experience and therefore the intelligent environment could be perceived as annoying.

4 Existing Standards

Some present standards (Lon Talk, KNX and Zigbee) are described in this section. Their most important features are also analyzed.

4.1 Lon Talk

Lon Talk [12] is the underlying protocol through which devices in a LonWorks network communicate. It was originally created by Echelon [15]. In January 2009, Lon Talk was ratified as a global standard for building controls, being now formally known as ISO/IEC 14908-1 [16].

The ISO/IEC 14908-1 protocol provides a set of services that lets a device's application program send and receive messages to and from other network devices, without the necessity to know the topology or the other devices' names, addresses, or functions. The protocol can optionally provide end-to-end acknowledgement of messages, authentication of messages, and priority delivery to provide bounded transaction times.

This protocol adapts better to the applications required by AmI environment, and in contrast with X10, it covers all layers of the OSI model. A simple scheme of its communication stack can be seen in Fig. 2.

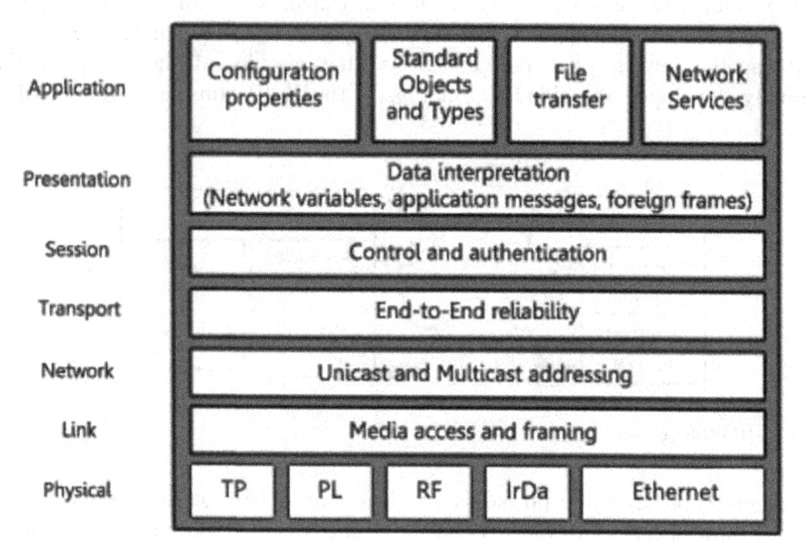

Fig. 2 Lon Talk protocol stack.

4.2 KNX

KNX is meant to be the successor to, and convergence of, three previous standards that are now deprecated: the European Home Systems protocol (EHS), BatiBUS and the European Installation Bus (EIB, EIBus or Instabus).

Those protocols are comprehensively described in the CLC/TR 50552:2010 technical report from the CENELEC [17]. They include the most common connection types such as twisted pair, radio frequency, infrared data association, power line connection and are compatible with the internet protocol. As it is shown in Fig. 3, the architecture of the KNX model includes several connections and up to three configuration modes for the AmI devices.

This standard allows for enabling access to the system via LAN, analog or mobile phone networks for having a central or distributed control of the system via PCs, touch screens and Smartphones.

KNX was approved as an International Standard [19] as well as a Canadian standard [20], so as an US Standard [21], a European Standard [14] and Chinese Standard (GB/Z 20965). It was not designed for environmental control such as lighting or heating and did not include any security issues at first. Extending KNX towards new application areas like access control or security systems leads to increasing demands. For this reason, an extension of this protocol, called EIBsec, was developed.

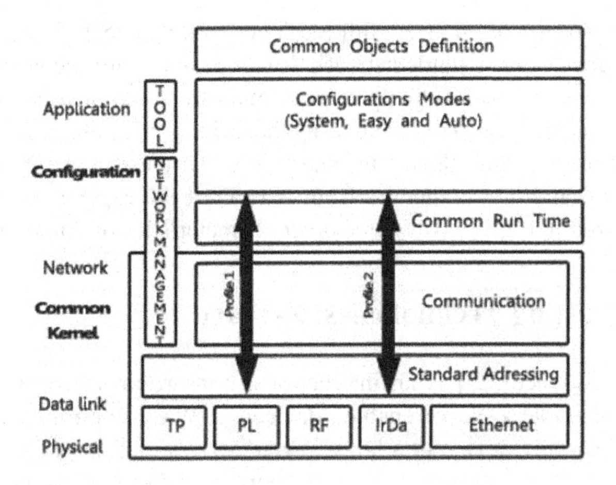

Fig. 3 The KNX model [18].

From the official information the KNX-Association currently has about 250 members from 29 different countries [22]. They also claim that Worldwide KNX Association comprises more than 30,000 partners in 75 countries [23].

4.3 Zigbee

This standard was developed as a very low-cost, low-power-consumption, two-way and wireless communications standard [24].

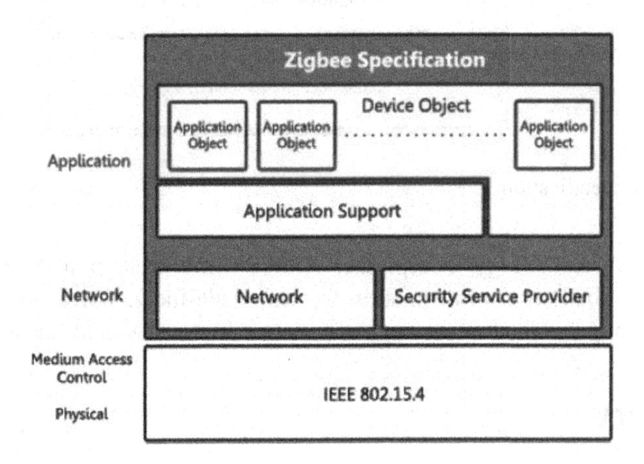

Fig. 4 The Zigbee model.

The Zigbee specification offers full mesh networking capabilities of supporting thousands of devices on a single network. It was designed for the hostile radio frequency environments that routinely exist in mainstream commercial applications. The provided security services include methods for key establishment and transport, frame protection and device managements. All these services are described in the chapter 4 and in the annexes from A to D of the Zigbee specification and can be visualized in Fig. 4, so as some other characteristics of this standard.

5 Other Existing Technologies: Z-Wave

The Z-Wave Alliance [25] is an international consortium of manufacturers that provide interoperable Z-Wave enabled devices. Z-Wave control can be added to almost any electronic device in a house, even devices that no one would think of as "intelligent". This technology was developed to cover all layers of the OSI stack, as shown in Fig. 5.

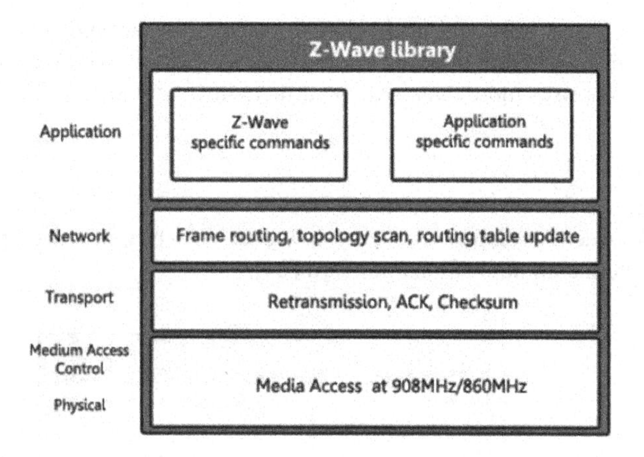

Fig. 5 Z-Wave specification.

The Z-Wave technology incorporates AES128 link protection. Most “secure” applications embed secure information frames inside the Z-Wave transport payload. Z-Wave appears in a broad range of products in the U.S. and Europe.

6 Summary

Key features of the analyzed communication solutions are summarized in Table 1 to ease comparison.

Table 1 Comparison of the main features analyzed.

Name	Type	Price	Connection	Last update	Security
X-10	Protocol	Free of charge	PL	Oct 2, 2011	Implemented
Lon Talk	Standard	Free, license submittal required	PL/TP	Nov 3, 2011	Implemented
KNX	Standard	6,34 €[1]	IP/PL/RF TP/IrDA[2]	Jun 17, 2010	Provided
Zigbee	Standard	Only required for commercial gain	RF	Oct 28, 2011	Implemented
Z-Wave	Other technology	Depending on the product	RF	Nov 2, 2011	Provided

7 Conclusions and Future Work

This comparative study tries to help inexperienced developers when deciding which technology best suits the requirements of a system to be implemented (whether it is AmI or not) and what the costs will approximately be.

According to the analyzed technologies, it may be the time to rethink and look at the potential of technologies such as IPv6. Thus, as future work this study proposes to analyze the suitability of such official standards in relation to AmI environments. It is also proposed to study the possibilities regarding the interconnection with the wide range of present standards (apart from the ones in this paper).

Other lines of future work could be: the creation of a modular platform to integrate different solutions, making end user applications simpler; the implementation of different agents in between the devices, making them AmI devices that intercommunicate and build up a Smart Environment.

References

[1] IBM Corp. The IBM vision of a smarter home. Somers, New York (2010)

[2] Turner, K.J.: Flexible management of smart homes. J. of Ambient Intell. and Smart Environ. (2011), doi: 10.3233/AIS-2011-0100

[3] Jakimovsky, P., Becker, F., Sigg, S., Schmidtke, S., Beigl, M.: Collective Communication for Dense Sensing Environments. In: 7th Int. Conf. in Intell. Environ. (2001), doi: 10.1109/IE.2011.42

[4] X10. X10 Knowledgebase Wiki (2011), http://kbase.x10.com/wiki/Main_Page

[1] Buying the EN 50090.

[2] PL (Power Line), RF (Radio Frequency: Bluetooth, Wi-Fi, etc.), TP (Twisted Pair) or IrDA (Infrared Data Association).

[5] Konnex Assoc. KNX (2011), http://www.knx.org (accessed January 9, 2012)
[6] Zigbee. Zigbee Alliance Web (2011), http://www.zigbee.org/ (accessed January 9, 2012)
[7] Lee, J.S., Su, Y.W., Shen, C.C.: A Comparative Study of Wireless Protocols: Bluetooth, UWB, ZigBee, and Wi-Fi IECON 2007. In: 33rd Annual Conference of the IEEE (2007), doi:10.1109/IECON.2007.4460126
[8] Lin, Y.J., Latchman, H., Newman, R., Katar, S.: A comparative performance study of wireless and power line networks. IEEE Communications Magazine (2003), doi:10.1109/MCOM.2003.1193975
[9] Salvador, Z., Jimeno, R., Lafuente, A., Larrea, M., Abascal, J.: Architectures for ubiquitous environments. In: Intern. Conf. on Wireless and Mobile Comput., Networking and Commun., WiMob 2005 (2005), doi:10.1109/WIMOB.2005.1512954
[10] Miori, V., Tarrini, L., Manca, M., Tolomei, G.: An open standard solution for domotic interoperability. IEEE Trans. on Consumer Electron. (2006), doi:10.1109/TCE.2006.1605032
[11] Edwards, W.K., Grinter, R.E.: At Home with Ubiquitous Computing: Seven Challenges. In: Abowd, G.D., Brumitt, B., Shafer, S. (eds.) UbiComp 2001. LNCS, vol. 2201, pp. 256–272. Springer, Heidelberg (2001)
[12] Echelon. Lon Talk Protocol (2011), http://www.echelon.com/support/documentation/manuals/general/078-0183-01B_Intro_to_LonWorks_Rev_2.pdf
[13] Z-Wave. Z-WaveStart (2011), http://www.z-wave.com/modules/ZwaveStart/
[14] Rye, D.: X10 - Home gadgets since 1978 (2008), ftp://ftp.x10.com/pub/manuals/xtdcode.pdf
[15] Echelon. Echelon Corp. (2011), http://www.echelon.com/ (accessed January 9, 2012)
[16] JTC 1. Intern. Organ. for Stand. (2011), http://www.iso.org/iso/iso_catalogue/catalogue_tc/catalogue_detail.htm?csnumber=60203
[17] CENELEC. Eur. Comitee for Eur. Stand. (1996), http://www.cenelec.eu/dyn/www/f?p=104:110:2512870545098707::::FSP_ORG_ID,FSP_LANG_ID,FSP_PROJECT:,25,4601
[18] Konnex Assoc. KNX Sys. Archit. Konnex Assoc., Brussels (2004)
[19] ISO/IEC. Intern. Organ. for Stand. (2007), http://www.iso.org/iso/iso_catalogue/catalogue_tc/catalogue_detail.htm?csnumber=43842
[20] CSA. CSA Stand (2007), http://shop.csa.ca/en/canada/information-technology/cancsa-isoiec-14543-3-1-07/invt/27027032007/
[21] Am. Soc. of Heat., Refrig. and Air-Cond. Eng. Inc. ASHRAE (2010), http://www.ashrae.org/technology/page/132
[22] KNX Assoc. KNX Member List (2008), http://www.knx.org/knx-members/list/
[23] KNX Assoc. KNX Partner List (2008), http://www.knx.org/knx-partners/knxeib-partners/list/
[24] TG4. IEEE 802.15 (2010), http://www.ieee802.org/15/pub/TG4.html
[25] Z-Wave Alliance. Z-Wave Alliance (2011), http://www.z-wavealliance.org/modules/AllianceStart/

Guardian: Electronic System Aimed at the Protection of Mistreated and At-risk People

Ricardo S. Alonso, Dante I. Tapia, Óscar García, Fabio Guevara,
José A. Pardo, Antonio J. Sánchez, and Juan M. Corchado

Abstract. Ambient Intelligence (AmI), based on ubiquitous computing, represents the most promising approach between people and technology to solve the challenge of developing strategies that allow the early detection and prevention of problems in automated dependence environments. One of the most influenced areas by AmI-based systems will be security and, more specifically, the protection of people under risk situations, including cases of mistreatment or loss. This will contribute to improve important aspects of the quality of life of these people, specially their safety. This paper describes Guardian, an integral solution designed for improving the protection of mistreated and at-risk people.

Keywords: Ambient Intelligence, Mistreatment, Safety, GPS, GPRS, ZigBee.

Introduction

According to statistical data from the Ministry of Interior of Spain [1], from 2006 to 2008 more than 60 women were killed in Spain each year by their couples or ex-couples. In 2008, a 23% of the killed women had filed a formal complaint against their murderers. In this regard, this paper describes *Guardian*, an Ambient

Ricardo S. Alonso · Dante I. Tapia · Óscar García · Fabio Guevara
R&D Department, Nebusens, S.L., Scientific Park of the University of Salamanca,
Edificio M2, Calle Adaja, s/n, 37185, Villamayor de la Armuña, Salamanca, Spain
e-mail: {ricardo.alonso,dante.tapia,oscar.garcia,
fabio.guevara}@nebusens.com

José A. Pardo · Antonio J. Sánchez · Juan M. Corchado
Department of Computer Science and Automation, University of Salamanca,
Plaza de la Merced, s/n, 37008, Salamanca, Spain
e-mail: {jose.pardo,anto,corchado}@usal.es

P. Novais et al. (Eds.): Ambient Intelligence - Software and Applications, AISC 153, pp. 11–18.
springerlink.com

Intelligence (AmI) [2] based electronic system aimed at the location and protection of people under risk situations. These risk situations include prevention of aggressions to threatened people, as well as surveillance and care of children, elderly and other vulnerable people. Therefore, the main aim is developing a system capable to protect people under risk of being mistreated, assaulted or lost, in a totally autonomous way by means of wireless electronic devices.

In this sense, Ambient Intelligence (AmI) is an emerging multidisciplinary area based on ubiquitous computing and that influences on the design of protocols, communications, systems, devices, etc. [2]. Ambient Intelligence proposes new ways of interaction between people and technology, making the latter to adapt to the users' needs and the environment that surrounds them [3]. This kind of interaction is reached by means of technology that is embedded, non-invasive and transparent for users, whose main aim is facilitating their daily activities [4]. An environment capable of recognizing the presence of people, and locating them in a geographical an activity context is the base to AmI to demonstrate all its potential.

In situations of violence against women exercised by husbands or couples, or in the framework of other emotional relations, the authorities consider the electronic surveillance as an indispensable tool for helping to guarantee the safety of victims. Thus, there are different approaches that propose electronic telemonitoring systems for tracking victims and aggressors [5] [6] to reduce risk situations.

The Guardian system will make use of different wireless technologies, such as A-GPS [7], GPRS [8] and ZigBee [9], to provide the majority of its features. Therefore, this project involves the design and creation of a completely distributed innovative hardware and software platform. This platform will have to exceed the available systems currently available in the market, integrating all its functionalities by means of a powerful logic of middleware layers.

The next section describes the problem of protecting potential victims of domestic violence, as well as some existing approaches that try to solve this problem. After that, the main components of the novel Guardian system are depicted. Finally, the conclusions and the future work are presented.

Background and Problem Description

It is important to point out that, in Spain, the pressure over the assistance system is growly increasing, as well as the necessity to offer services with more quality. Thousands of people (mainly women) are daily mistreated, battered and abused by their ex-couples or their current couples [1]. These mistreated people suffer from a lack of freedom and the violation of their most elemental rights.

This fact represents a complex challenge, which coincides with a crisis in the support systems that try to provide solutions to these necessities. In this scenario, the technology can play a decisive role. Initiatives such as the Guardian project, oriented to improve the assistance services for these population segments, involve a strategic relevance. Although the current number of incidents related to violence against women exercised by husbands or couples, or in the framework of other emotional relations, are increasingly more advertised than in the past [1], there are still hidden facts to be known. In such situations, one of the most important aspects is assuring that the potential aggressor and victim are physically separated by a certain safety distance. In this sense, there are several approaches developed during the last years with different features that try to solve the problem of locating and detecting the proximity between aggressor and victim.

There are approaches centered on monitoring and locating accurately the aggressor, such as the BI ExacuTrack® One system [5]. This system consists of a light, resistant and tamper-proof device that is worn by the aggressor on its ankle, offering long battery autonomy. The locating process is performed using a combination of several technologies, including autonomous GPS, A-GPS [7] and AFLT (Advanced Forward Link Trilateration) [10]. Using this combination of wireless technologies, the system can estimate accurately the position of the user, even in hard conditions, such as indoors, vehicles in motion or between high buildings. Nevertheless, in this system the energy consumption and the need of battery recharge are very exigent.

Other approach is the One Piece GPS System [6], whose main objectives are the device ubiquity, less necessity of device maintenance and robustness against tries of manipulation on the aggressor's device. This system includes an integrated active GPS device [7] that combines a GPS receiver, a microprocessor and different communication components in a wrist or ankle bracelet. In situations where the aggressor's device is close to restricted areas, the device sends notifications to the supervisor agents in real time through fax, email or SMS. Supervisor agents can create inclusion or exclusion areas that surround a specific geographic location, as the victim's residence or work place. This implies an increase in the total cost of the system, as it requires a management of the electronic borders by the supervisor agents, as well as a previous learning process.

Considering the limitations of the electronic borders, Omnilink [11] proposes a dynamic tracking system which calculates the distance between victim and aggressor in real time. Its monitoring system allows agents to control the movements of the aggressor according to the movements of its victim, both indoors and outdoors. Using a combination of monitoring devices in victim and aggressor, the agents can control the proximity between both. If the aggressor is too close to the victim's home, work place or some of the exclusion areas, both the agents and the victim are notified. In addition, to support the electronic borders management, this solution allows calculating the proximity between victim and aggressor in real time and acting consequently through specific software in the data center. Although this proximity calculation represents an improvement over other systems, it is important to take into account that this calculation can be conditioned by possible congestions in the communication network or by low radio coverage according to the positions of victim and aggressor. In this sense, the Guardian system solves this limitation as the calculation of proximity between victim and aggressor is not delegated to other components of the system. Thus, the own tracking devices are the ones responsible for detecting each other at a certain distance and calculating it from the parameters of the received radiofrequency signals.

Nevertheless, these approaches do not cover completely indoor situations where GPS or GPRS coverage can fail or such approaches need to define and manage exclusion areas. In this sense, the Guardian system includes devices that integrate GPS, GPRS and ZigBee in the same wireless module, thus covering all situations in an autonomous way.

The Guardian System

This section describes the main components of the Guardian system, whose basic schema is shown in Figure 1. First, the basic functioning of the system is depicted. After that, the different hardware modules that make up each wireless device are described. As this is a research work that will be finished in Q4 2012, this paper presents a preliminary description that will be extended and published further on.

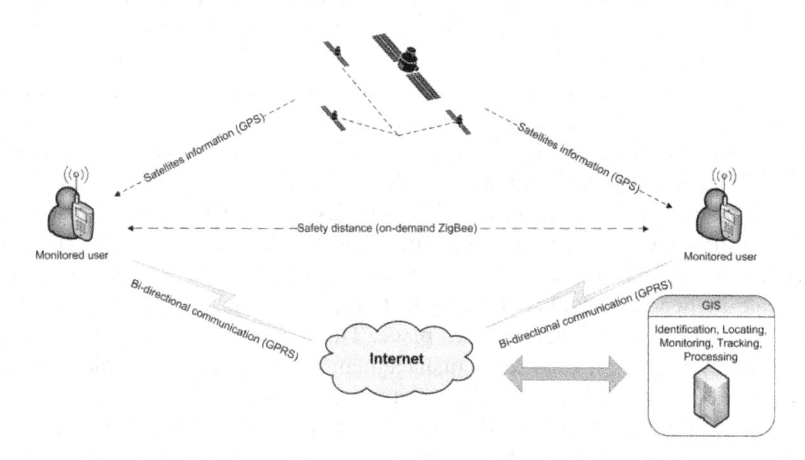

Fig. 1 Basic schema of the Guardian system.

Figure 2 shows the flow diagram representing the basic functioning of the system at the wireless infrastructure level. The basic functioning of the Guardian system is as follows. There are two kinds of users or roles in the system. On the one hand, the first kind of user, the threatened or at-risk user, carries or wears a mobile phone or other mobile device (e.g., a bracelet). This device includes A-GPS, GPRS and ZigBee capabilities. On the other hand, the other kind of user (i.e., the potential aggressor) also wears another device with similar wireless capabilities.

Both devices (the victim's one and the potential aggressor's one) obtain their position making use of their A-GPS module. This way, both devices send its position using GPRS to the control center where the Geographic Information System (GIS) is running. Therefore, the control center keeps track of the positions of both users. However, this information is not enough to achieve an efficient protection. This is because the GPS or A-GPS technologies do not properly work on some situations, as indoor locations (e.g., buildings or tunnels) [7]. Is in this point where the Guardian system makes the difference over the conventional systems currently available in the market.

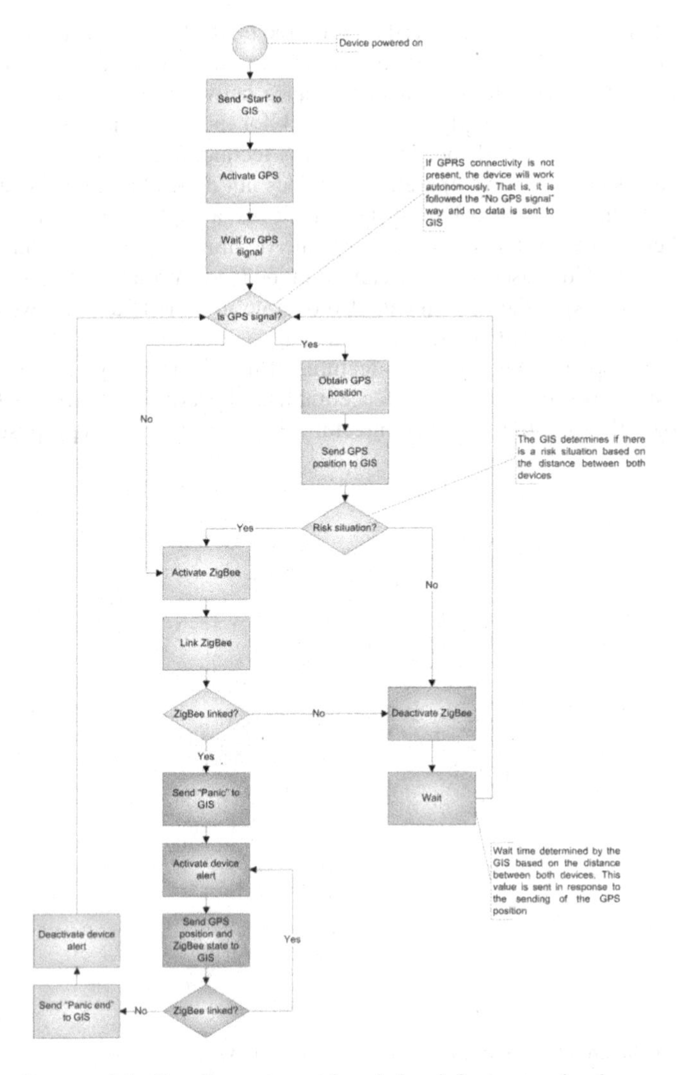

Fig. 2 Flow diagram of the Guardian system at the wireless infrastructure level.

The ZigBee technology [9] covers those situations where GPS or A-GPS cannot work correctly. In the Guardian system, the ZigBee module is used when users are nearly located. In addition, ZigBee is activated when there is no A-GPS or GPRS coverage. At this moment, both devices (the victim's one and the potential aggressor's one) start searching the signal from the counterpart device. The transmission power of both ZigBee modules can be selected by software. According to this transmission power and the sensitivity of the antennas, the ZigBee signal range can reach even several kilometers. Therefore, if one of the devices detects the other one, it sends an alert to the control center using GPRS. If there is no GPRS coverage, both devices will raise visual and acoustic alarms, using the buzzers included in the devices. Furthermore, the control center keeps track of the last positions of both devices before the A-GPS or GPRS coverage was lost. This way, the control center determine if the distance between users and the last time before losing the coverage imply a potential risk for the threatened user.

Using the combination of the three wireless technologies (A-GPS, GPRS and ZigBee) the system is always operative and does not depend on an only technology to work. Thus, the Guardian system achieves a higher level of autonomy against other similar systems. Furthermore, the flexibility of the Guardian system allows that one of the used devices can be embedded into an object, such as an access door. This way, the system can also operate in a mode that allows controlling the access to protected areas.

In order to implement many of the features of the Guardian system, it is necessary to design specific devices that accomplish the criteria established in the project. To do that, the functional architecture shown in Figure 3 is proposed.

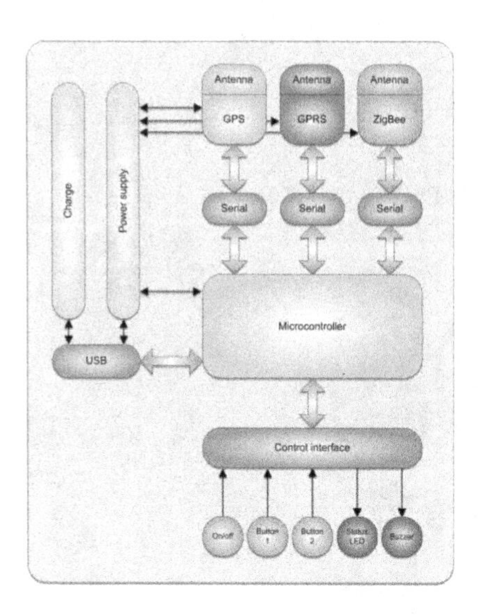

Fig. 3 Functional architecture of the Guardian system's devices.

This functional architecture is based on the n-Core® Sirius B devices from Nebusens [12]. Each n-Core® Sirius B device includes an 8-bit RISC (Atmel ATmega 1281) microcontroller with 8KB of RAM, 4KB of EEPROM and 128KB of Flash memory and an ZigBee transceiver. n-Core® Sirius B devices have both 2.4GHz and 868/915MHz versions and have several internal and external communication ports (GPIO, ADC, I2C, PWM, and USB/RS-232 UART) to connect to distinct devices, including a wide range of sensors and actuators. n-Core® Sirius devices form part of the n-Core platform [12], which offers a complete API (Application Programming Interface) to access all its functionalities. This platform was chosen because it provides all the needed functionalities by means of its full-featured n-Core firmware and the n-Core API. Thus, developers do not have to write any additional embedded code to build a system as this, but just configure the n-Core Sirius devices to accomplish the required features [12].

The wireless devices used in the Guardian system are formed by different functional units or hardware modules interconnected to each other. Next, the main features of each of the units are briefly described.

- *Charge and power supply*: consists of a portable power supply system, that is, a rechargeable battery and a charge and power supply system. The charge system will be implemented in the same module and will allow charging the battery through a USB port.
- *USB*: communication interface between the device and a personal computer. On the one hand, it will allow configuring the device. On the other hand, it will allow charging the internal battery.
- *GPS*: this is the module responsible for obtaining the coordinates with the device position and offering them to the microcontroller through a serial data interface.
- *GPRS*: this module facilitates the communication between the device and the control center, making use of the TCP/IP protocol over GPRS for the information transmission.
- *ZigBee*: this is the module responsible for calculating the relative distance between two paired devices. This module can work when there is no GPS coverage or as additional information to that provided by the GPS module (e.g., in situation of imminent proximity).
- *Microcontroller*: this is the core of the device and is responsible for processing the information received from the different functional units, as well as giving response to them to coordinate their correct functioning. Amongst its multiple features we have the battery usage optimization.
- *On/off*. This button allows powering *on* or *off* the device.
- *Button 1*: When this button is pressed, the device sends a "Panic" alert to the control center.
- *Button 2*: General purpose button. This button can be configured from the control center to offer specific on-demand features.
- *Status LED*: This element is made up of two Light Emitter Diodes. These LEDs will show the distinct status of the devices, such as a low battery warning, an alert situation or a network failure, amongst others.
- *Buzzer*: an acoustic indicator that alerts the user when a situation requires its attention, such as an alert or panic situation.

Conclusions and Future Work

The Guardian system pursuits a revolutionary concept: the total supervision of people under risk situations, augmenting their safety and autonomy in a completely ubiquitous way. It is important to mention that there is no similar solution in the market, specifically a device with the characteristics specific to develop this project. This fact implies a high level of hardware development. Furthermore, there is not a hardware/software platform that fully provides the middleware layers necessary to integrate all the mentioned technologies.

Future work includes the full development of all the projected functionalities. This includes the production of the first hardware prototype for the wireless devices. At the software level, both the firmware embedded in the devices and middleware/software layers at the GIS will be completely developed, integrated and configured. Then, the system will be implemented in a simulated scenario to test if it is suitable for the situations for which it is designed. These situations include domestic violence and children/elderly care situations. Finally, the system will be implemented on a real scenario in order to test its actual performance.

Acknowledgments. This project has been supported by the Spanish Ministry of Science and Innovation. Project IPT-430000-2010-035.

References

1. Ministry of Interior of Spain, Violencia de género. Programa de Intervención para Agresores (2010), http://www.mir.es/file/53/53009/53009.pdf (accessed December 31, 2011)
2. Aarts, E., de Ruyter, B.: New research perspectives on Ambient Intelligence. Journal of Ambient Intelligence and Smart Environments 1(1), 5–14 (2009)
3. Bajo, J., de Paz, J.F., de Paz, Y., Corchado, J.M.: Integrating case-based planning and RPTW neural networks to construct an intelligent environment for health care. Expert Syst. Appl. 36, 5844–5858 (2009)
4. Tapia, D.I., Abraham, A., Corchado, J.M., Alonso, R.S.: Agents and ambient intelligence: case studies. Journal of Ambient Intelligence and Humanized Computing 1(2), 85–93 (2010)
5. BI Incorporated. One-piece active GPS offender tracking: BI ExacuTrack® One (2011), http://bi.com/exacutrackone (accessed September 9, 2011)
6. iSECUREtrac. One-piece GPS Systems from iSECUREtrac (2011), http://www.isecuretrac.com/Services.aspx?p=GPS#onepiece (accessed September 9, 2011)
7. Djuknic, G.M., Richton, R.E.: Geolocation and Assisted GPS. Computer 34(2), 123–125 (2001)
8. Haung, Y.-R., Lin, Y.-B.: A bandwidth-on-demand strategy for GPRS. IEEE Transactions on Wireless Communications 4(4), 1294–1399 (2005)
9. Baronti, P., Pillai, P., Chook, V.W.C., Chessa, S., Gotta, A., Hu, Y.F.: Wireless sensor networks: A survey on the state of the art and the 802.15.4 and ZigBee standards. Comput. Commun. 30(7), 1655–1695 (2007)
10. Küpper, A.: Location-Based Services: Fundamentals and Operation, 1st edn. Wiley (2005)
11. Anynye, S., Rajala, Y.: System and Method for Tracking, Monitoring, Collecting, Reporting and Communicating with the Movement of Individuals. U.S. Patent Application, 20100222073 (February 9, 2010)
12. Nebusens: n-Core®: A faster and easier way to create Wireless Sensor Networks (2012), http://www.n-core.info (accessed January 2, 2012)

Sensor-Driven Intelligent Ambient Agenda

Ângelo Costa, José Carlos Castillo Montotya, Paulo Novais,
Antonio Fernández-Caballero, and María Teresa López Bonal

Abstract. The rapid increasing of elderly persons living alone presents a need for continuous monitoring to detect hazardous situations. Besides, in some cases, the forgetfulness of these people leads to disorientations or accidents. For these reasons, this paper introduces a system that brings together visual monitoring capabilities to track people and to perform activity detection, such as falls, with scheduling abilities to intelligently schedule urgent and non-urgent events.

Keywords: Monitoring System, Activity Detection, Cognitive Assistants, Memory Enablers, Personal Assistants, Body Area Networks.

1 Introduction

Projections have displayed the rapid increase of elderly persons needing help to execute daily tasks [1]. The growth of the elderly community means that each year the rate of the problems and the persons that have to be treated is multiplied. The economic cost involved maintaining a person in a hospital or a healthcare facility, under constant human monitoring, also greatly grows [2]. But it is also quite problematic to leave a person in need alone in his/her house. Events such as falls tend to occur frequently when persons are older, and the problems that arise from these falls are very serious, since more than hematomas or broken bones use to occur. Indeed, they may be underlying problems that can possibly be undetected, turning later on into severe health problems [3]. Another frequent situation consists in the

Ângelo Costa · Paulo Novais
Department of Informatics/CCTC
University of Minho, Braga, Portugal
e-mail: {acosta,pjon}@di.uminho.pt

José Carlos Castillo Montotya · Antonio Fernández-Caballero · María Teresa López Bonal
Instituto de Investigación en Informática de Albacete (I3A), n&aIS Group,
University of Castilla-La Mancha, Albacete, Spain
e-mail: {JoseCarlos.Castillo,Antonio.Fdez,Maria.LBonal}@uclm.es

P. Novais et al. (Eds.): Ambient Intelligence - Software and Applications, AISC 153, pp. 19–26.
springerlink.com

forgetfulness of elderly persons, being the loss of memory of recent events somehow frequent. For instance, forgetting that they fell is quite common, adding the fact that they tend not to call for help in most of the cases, leading to new later complications [4,5]. Also, falls indicate deeper problems, such as loss of cognitive capabilities or underlying cerebral problems. But there is still the worst scenario: the fallen person cannot move or even call for help, being unable to perform any action. In this case, computer systems are the perfect solution to provide help in a non-intrusive way, taking proactive actions on behalf of the user by choosing the right kind of help required.

This paper introduces a system that brings together visual monitoring capabilities to track people and to perform activity detection, such as falls, with scheduling abilities to intelligently schedule urgent and non-urgent events according to the sensed data. The paper is organized as follows. Section 2 presents the system overview, especially paying attention to the interaction between the system and the external elements as well as describing the information sources proposed for the monitoring purpose. Section 3 presents the lower level modules, in charge of monitoring and simple event detection whilst in section 4 the higher level monitoring through activity detection is presented. In section 5 the scheduling and decision-making processes are described. Finally, some conclusions are presented.

2 System Overview

The proposed system consists of several processing modules that work with different information abstraction levels in order to manage a user's agenda by adding new events according to the information sensed from the environment. For this purpose, a series of sensors are placed in the monitored scenario and attached to the user. The former series of sensors are not only visual but there are also other kinds of sensors (volumetric and movement detectors, audio, etc.), available in the commercial field [6,7]. The latter are an implementation of a Body Area Network, being a portable sensor network with discrete sensors that read vital signs and, with the help of the clinical guidelines, provide proactive measures if any problem or any deviation of the pattern is detected.

2.1 System Modules

The system modules can be divided into three categories, namely, data acquisition & low level processing, high level processing, and scheduling & decision making (see Fig. 1). The first category involves those modules directly connected to the sensors, performing data acquisition, its preprocessing and the first information processing. Sensors are clustered (e.g. sensors in the same room) and each cluster is processed by a different module. The distribution of the processing modules of the sensors also provides scalability (new sensors and processing nodes can be added without changes in the previous modules) and fault tolerance (a crashing module does not compromise the overall performance) to the system. On the other hand, information reaches a higher level of abstraction with the "*activity detection*" module. This higher level module gathers information from the previous ones, generating a complete view of the monitored environment that enables the detection of global activities. Finally, activities information is utilized by the scheduling & decision making module in combination with the user preferences and clinical guidelines to update the profile of the user agenda and, should it be necessary, to trigger alarms for warning the emergency services.

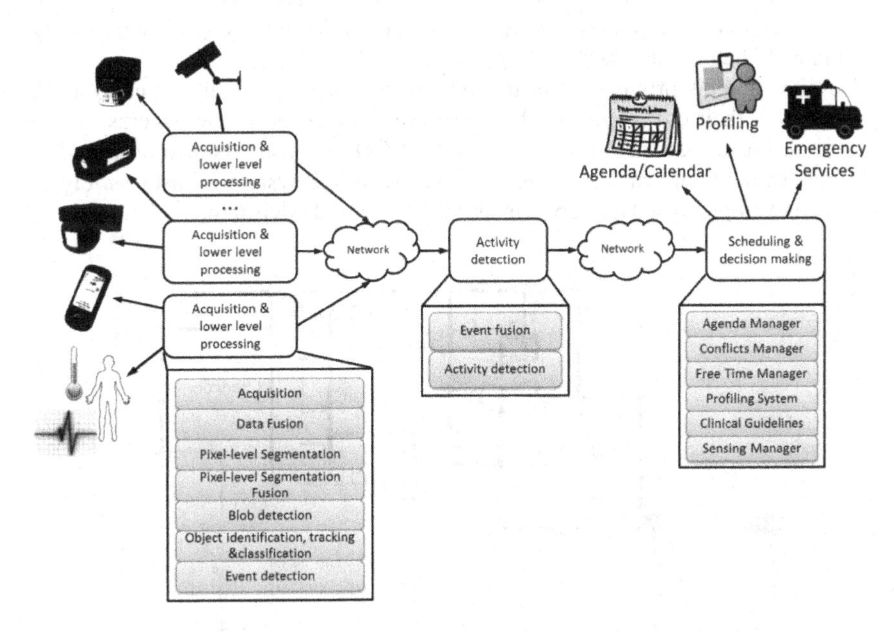

Fig. 1. System overview. Interaction with external elements and processing layers

The system described requires high computation resources to execute each level's algorithms. For this reason, a hybrid execution model is proposed to balance the system operation [8]. Data collection and the first levels of the processing stack are performed by "data acquisition & low level processing" nodes whilst a "high level processing" node is in charge of the collection of remote nodes data, of their fusion and of executing higher processing levels. Finally a "scheduling & decision making" node manages the user's agenda. It also manages the user's preferences and it continuously checks information from lower level nodes, considering the detected actions carried out by the user with the purpose of acting in consequence (e.g. through warning the emergency services when a fall is detected).

2.2 *Information Sources*

Inputs to the system are given by a set of sensors that, connected to the remote processing nodes, capture information of the scenario they are placed in.

These sensors are divided into two categories: visual and non-visual. Regarding the visual sensors used by the system, two technologies are employed. Firstly, traditional cameras are used in a wide range of application scenarios such as surveillance, sports monitoring, or user interaction, among others [9,10]. These cameras cover most of the monitored environment keeping the number of devices low. Considering the monitoring in homes, certain areas, such as bedrooms or toilets, are especially unsuitable for traditional image capture. That is the reason why thermal-infrared cameras are also considered. These cameras, which capture in low illumination conditions or through smoke or fog, keep human privacy by only obtaining the heat print of the persons.

Despite cameras provide great amounts of information, certain scenario areas can be out of their coverage. Besides, there are some areas where cameras cannot be placed. For these reasons, the utilization of COTS sensors is also proposed to reinforce cameras information, this way improving the system robustness. Fig. 2 shows an example of sensor deployment in a monitored environment.

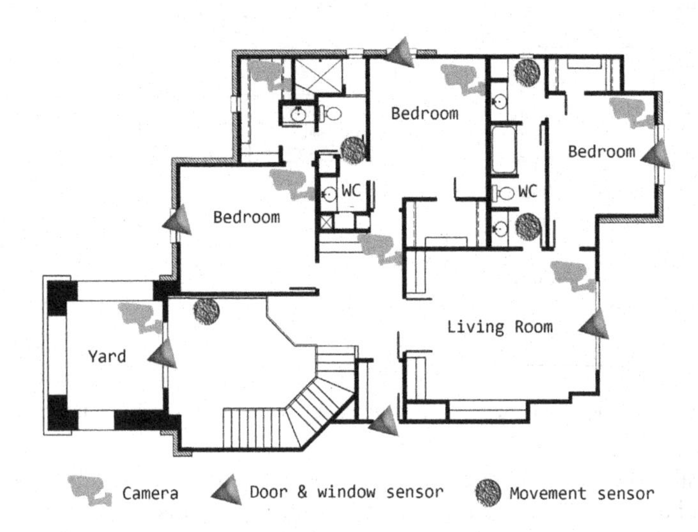

Fig. 2 Sensor placement in a test environment

3 Data Acquisition and Low-Level Processing

Low level operation can be divided into a set of processing levels that, operating in cascade, are able to detect humans and to identify some simple behaviors (events). As the system is designed to work with several sensors at the same time, operation at this level is distributed through modules. Each one is in charge of the acquisition and processing of a set of sensors following the traditional monitoring systems stack identified in [8]. These nodes send their information to a central high-level node that merges the information. Next, the different levels of *"acquisition and low level processing"* nodes are described (see Fig. 1).

- **Acquisition** interacts with the digital analog devices, measuring from the physical world and adapting these measures to be usable by the system. The acquisition level also performs information preprocessing.
- **Sensor Fusion** merges sensor data to improve the information quality (more complete and accurate). It introduces knowledge on the domain.
- **Pixel-Level Segmentation** isolates and filters the objects of interest contained in the input images. This level may hold a wide range of methods, such as binarization [11] or motion computation-based approaches [12].
- **Pixel-Level Segmentation Fusion** fuses images obtained ate the pixel-level segmentation stage as there might be several approaches running in the system (e.g. one devoted to color images and another one to infrared images).
- **Blob Detection** extracts information associated to the spots allowing a more efficient analysis of the objects.
- **Object Identification, Tracking & Classification** enhances the information abstraction, considering the temporal component to create objects from the blobs and mapping object coordinates from image coordinates system into real world coordinates. Also, it provides knowledge about "what" the objects are and their orientation, and calculates the trajectories followed by the moving objects within the scenario, making predictions about their future positions.
- **Event Detection** generates semantic information related to the behavior of the objects present in the scenario. Some examples are events such as running, walking or falling, which can be detected with just one or at most a few images.

4 High-Level Processing

• High-level processing is essential to obtain a global view of what is happening in the environment. Besides, in a multisensory system, where several nodes monitor a common scenario, the events generated usually do not match in the different sensors. For this purpose, two levels operate in this kind of node.

- **Event Fusion** unifies the information arriving from the different data generated in the previous lower-level nodes.
- **Activity Detection** analyzes and detects high level activities already associated to temporal features. After Event Fusion, the current level has a better knowledge of what is happening in the scenario according to the detected events. Hence, the activities detected at this level can be translated into actions along the scenario, providing a higher abstraction level.

After the execution of the previous levels, high-level information about the activities of the objects in the scenario is obtained. Next, this information is sent to the *scheduling and decision making* node that is in charge of decision-making and planning. Fig. 1 shows an overview of the system proposed so far, including the tasks associated to the last system node.

5 Scheduling and Decision Making

The scheduling and decision making node is in charge of the scheduling of events in the user agenda. This node works in two different paradigms: short term and long term actions. The short term actions are responsible of immediate and urgent events, such as calls to the emergency services or events that have to be tackled in the next few minutes. The long term actions are the normal events that have to be attended in the next hours or days. In layman's terms this node works as a scheduler and emergency manager, being able of receiving inputs of applications and sensors and working accordingly to the internal definitions of the system, choosing the most appropriate action and proceeding to it [13,14]. For this purpose, the module is designed following the multi-agent paradigm.

This module also carries out proactive actions recurring to agents that can analyze the high level input information. For it, different agents provide small decisions that can be merged later on, since every agent is responsible for a different area and different data. Next, the most relevant agents are described. Nevertheless, Fig. 3 shows the agents of the overall modules.

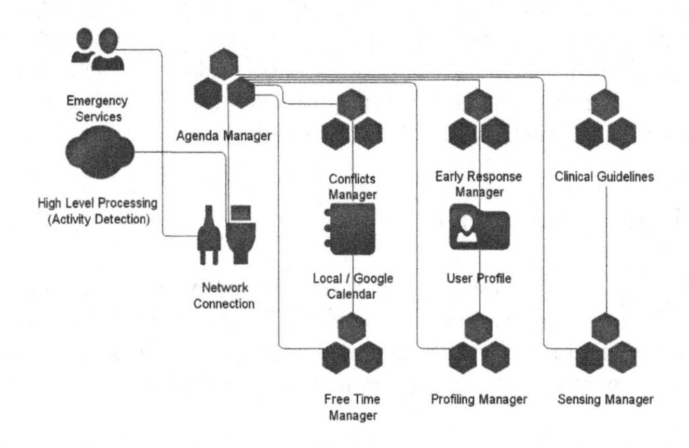

Fig. 3 Project architecture with the fall detection process highlighted

- **Agenda Manager** controls all agents, being responsible of starting them and delivering the information to the right agent.
- **Conflicts Manager** controls the incoming events from the previous module, and sorts them using a hierarchical system of importance to manage the overlapping of events, being able to move or delete events if needed.
- **Free Time Manager** controls the scheduling of leisure activities that stimulate the user. It schedules based upon a profiling system that contains the activities that the user likes and that are most appropriate.
- **Profiling Manager** uses Case Based Reasoning and Bayesian Networks to collect and improve the standard user profile over time, adjusting it better to the user's current preferences and other external conditionings.

- **Clinical Guidelines** consist of computer interpretable guidelines that provide a database to help to face a health problem by following strict functions of relational decisions.
- **Sensing Manager** is an implementation of a Body Area Network, consisting of a portable sensor network with discrete sensors that read vital signs. It shares the information with the agents present in the platform.

The customization of the "*high level processing*" module to detect falling in combination to an early alarm agent or a logging agent results on the creation of an "*Early response manager*". This agent agent is able to make intelligent decisions such as calling immediately the emergency services or analyzing the overall status and taking proactive measures, such as scheduling an appointment with a doctor or a social assistant. Adding the context of the ambient surrounding the user means that events involving the user, and actions regarding the sensed context can be calculated to prevent further problems and act over the present ones. The "*early response manager*" agent is continuously checking inputs from the "*high level processing module*", verifying if a fall has been detected. Thereon the system provides information, in form of data structures that contain post-processed information about the fall, valuable data such as intensity, post fall responsiveness, time and place, among other data, so the scheduling & decision making module can formulate knowledge and act accordingly. By using the Network Connection, messages are sent to user's smartphone asking if the user is injured or not, and finally calling the emergency service if there is no response after a period of time.

6 Conclusions and Future Work

This paper has introduced an approach for quick response in case of fallings or anomalous situations detected in an environment inhabited by elderly. This results in two endings: improved safety of the user and less hospital congestion. Keeping the user safe is a social demand for an active and happy aging, providing the technology needed for current and future generations.

For this purpose, three architectural blocks are presented. The first one is devoted to the detection and tracking of humans through a series of visual and non-visual sensors, as well as the detection of simple events. The second block fuses the simple events to provide a global view of the environment, being able to detect high level activities. Finally, the last block is in charge of the scheduling of events in the user agenda as well as the taking of actions according to the sensed data, with the purpose of reminding to take the medication or calling the emergency services. The use of these three blocks where visual and non-visual sensors interact and high-level information about the environment is obtained provides great advantages to avoid misdetections in contrast to other approaches such as [6,7]. In the future the development of the interoperability and the optimization of the connections between the different concepts and blocks and levels is foreseen.

Acknowledgments. This work is partially supported by the Spanish Ministerio de Ciencia e Innovación / FEDER under project TIN2010-20845-C03-01 and by the Spanish Junta de Comunidades de Castilla-La Mancha / FEDER under projects PII2I09-0069-0994 and PEII09-0054-9581. Also in collaboration with the AAL4ALL QREN 13852 project.

References

1. Beard, J.: A global perspective on population ageing. European Geriatric Medicine 1(4), 205–206 (2010)
2. Ageing, W.P. World Population Ageing 1950-2050. Population English Edition 26(26), 11–13 (2002)
3. S, A. World Population Ageing. Population English Edition 7(4), 750 (2009)
4. McEntee, W.J., Crook, T.H.: Age-associated memory impairment: a role for catecholamines. Neurology 40(3), Pt 1, 526–530 (1990)
5. Barker, A., Jones, R., Jennison, C.: A prevalence study of age-associated memory impairment. The British Journal of Psychiatry 167(5), 642–648 (1995)
6. Felisberto, F., Moreira, N., Marcelino, I., Fdez-Riverola, F., Pereira, A.: Elder Care's Fall Detection System. In: Novais, P., Preuveneers, D., Corchado, J.M. (eds.) ISAmI 2011. AISC, vol. 92, pp. 85–92. Springer, Heidelberg (2011)
7. Martín, P., Sánchez, M., Álvarez, L., Alonso, V., Bajo, J.: Multi-Agent System for Detecting Elderly People Falls through Mobile Devices. In: Novais, P., Preuveneers, D., Corchado, J.M. (eds.) ISAmI 2011. AISC, vol. 92, pp. 93–99. Springer, Heidelberg (2011)
8. Castillo, J.C., Rivas-Casado, A., Fernández-Caballero, A., López, M.T., Martínez-Tomás, R.: A Multisensory Monitoring and Interpretation Framework Based on the Model–View–Controller Paradigm. In: Ferrández, J.M., Álvarez Sánchez, J.R., de la Paz, F., Toledo, F.J. (eds.) IWINAC 2011, Part I. LNCS, vol. 6686, pp. 441–450. Springer, Heidelberg (2011)
9. Xu, M., Orwell, J., Lowey, L., Thirde, D.: Architecture and algorithms for tracking football players with multiple cameras. IEE Proceedings of Vision, Image and Signal Processing 152(2), 232–241 (2005)
10. Yao, Y., Chen, C.-H., Koschan, A., Abidi, M.: Adaptive online camera co-ordination for multi-camera multi-target surveillance. Computer Vision and Image Understanding 114(4), 463–474 (2010)
11. Fernández-Caballero, A., Castillo, J.C., Serrano-Cuerda, J., Maldonado-Bascón, S.: Real-time human segmentation in infrared videos. Expert Systems with Applications 38(3), 2577–2584 (2011)
12. Moreno-Garcia, J., Rodriguez-Benitez, L., Fernández-Caballero, A., López, M.T.: Video sequence motion tracking by fuzzification techniques. Applied Soft Computing 10(1), 318–331 (2010)
13. Costa, Â., Novais, P.: An Intelligent Multi-Agent Memory Assistant. In: Bos, L., Goldschmidt, L., Verhenneman, G., Yogesan, K. (eds.) Handbook of Digital Homecare - Successes and Failures, vol. 1, pp. 197–221. Springer, Heidelberg (2011)
14. Costa, Â., Novais, P., Corchado, J.M., Neves, J.: Increased performance and better patient attendance in an hospital with the use of smart agendas. Logic Journal of IGPL (2011), doi:10.1093/jigpal/jzr021

A Layered Learning Approach to 3D Multimodal People Detection Using Low-Cost Sensors in a Mobile Robot

Loreto Susperregi, Basilio Sierra, Jose María Martínez-Otzeta, Elena Lazkano, and Ander Ansuategui

Abstract. In this paper we propose a novel approach for low cost multimodal detection of humans with mobile service robots. Detecting people is a key capability for robots that operate in populated environments. The main objective of this article is to illustrate the implementation of machine learning paradigms with computer vision techniques to improve human detection using 3D vision and a thermal sensor. Experimental results carried out in a manufacturing shop-floor show that the percentage of wrong classified using only Kinect is drastically reduced with the classification algorithms and with the combination of the three information sources.

Keywords: Computer Vision, Machine Learning, Robotics, 3D People Detection, Multimodal People Detection.

1 Introduction

The introduction of service robots in environments such as museums, hospitals or at home, demand for robots that can adapt to complex and unstructured environments and interact with humans. Non expert users or even users with physical disabilities will share their space with robots and the robots therefore have to be able to determine human positions in real-time. Within this article, the objective of the mobile robot is to approach the closer person in the room, i.e. to approach the person to a given distance and to verbally interact with him. This "engaging" behaviour may be useful in potential robot services such a tour guide, health care or information provider. To accomplish this the robot must be able to detect human presence in its

Loreto Susperregi · Jose María Martínez-Otzeta · Ander Ansuategui
TEKNIKER-IK4, Spain
e-mail: lsusperregi@tekniker.es
jmmartinez@tekniker.es
aansuategui@tekniker.es

Basilio Sierra · Elena Lazkano
University of Basque Country, Spain
e-mail: b.sierra@ehu.es
e.lazkano@ehu.es

P. Novais et al. (Eds.): Ambient Intelligence - Software and Applications, AISC 153, pp. 27–33.
springerlink.com

vicinity and it cannot be assumed that the person faces the direction of the robot since the robot acts proactively.

This article describes the realization of a human detection system based on two low-cost sensing devices: Kinect motion sensor device, [2], and Heimann HTPA thermal sensor, [1]. Kinect offers a rich data set with some limitations, mainly Kinect rely on the detection of human activities captured by a colour vision static camera. To cope with this, the combination with thermal vision can help to overcome some of the problems as humans have a distinctive thermal profile, despite thermal sensors are very sensitive to heat sources, the sensor data does not depend on light conditions and people can also be detected in complete darkness.

This article outlines the design and development of a multimodal human detection system. The chosen approach is to combine machine learning paradigms with computer vision techniques. We have experimented in a real manufacturing shop-floor where machines and humans share the space in performing production activities. Experiments seem promising considering that the percentage of wrong classified using only Kinect detection algorithms is drastically reduced.

The rest of the paper is organized as follows: in section 2 related work in the area of human detection is presented. Section 3 describes the proposed approach and section 4 the experimental setup. Section 5 shows experimental results and Section 6 conclusions and future work.

2 Related Work

People's detection and tracking systems have been studied extensively. As a complete review on people detection is beyond the scope of this work, an extensive work can be found in [9, 4], we focus on most related to our work. To our knowledge, two approaches are commonly used for detecting people on a mobile robot. By the one hand, vision based techniques, and on the other hand, combining vision with other modalities, normally range sensors such as laser scanners or sonars.

Apart from cameras, the most common devices used for people tracking are laser sensors. One of the most popular approaches in this context is to extract legs by the detecting moving blobs that appear as local minimal in the range image. [7] presents a system for detecting legs and follow a person only with laser readings. Other implementations such as [3] also use a combination of face detection and laser-based leg detection.

Most existing combined vision-thermal based methods, in [6, 12], concern non-mobile applications in video monitoring applications. Some works, [5], show the advantages of using thermal images for face detection.

As yet, however, there is hardly any published work on using thermal sensor information to detect humans on mobile robots. The main reason for the limited number of applications using thermal vision so far is probably the relatively high price of this sensor. [11] shows the use of thermal sensors and greyscale images to detect people in a mobile robot. A drawback of most of these approaches is the sequential integration of the sensory cues.

Most of the abovementioned approaches have mostly used predefined body model features for the detection of people. Few works consider the application of learning techniques. [8] builds classifiers able to detect a particular body part such as a head, an upper body or a leg using laser data.

3 Proposed Approach

We propose a multimodal approach, which can be characterized by the fact that all used sensory cues are concurrently processed. The proposed detection system is based on a Kinect motion sensor device and a HTPA thermal sensor developed by Heimann, mounted on top of a RMP Segway mobile platform, which is shown in Figure 1. Kinect provides 3D images: a 640x480 distance (depth) map and a 640x480 RGB image in real time (30 fps). The HTPA sensor offers a 32x31 image that allows a rough resolution of the temperature of the environment.

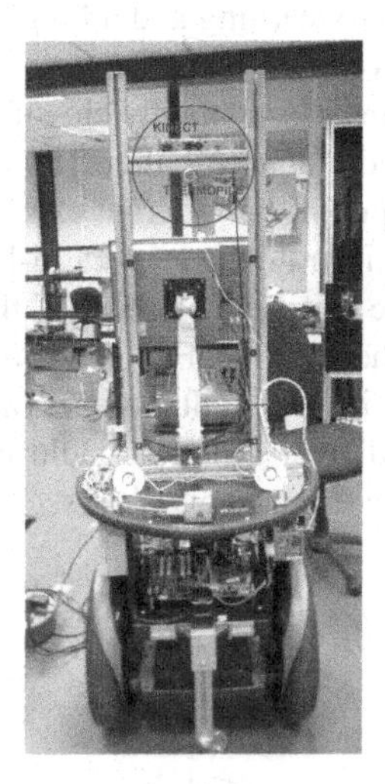

Fig. 1 The used robotic platform: a Segway RMP 200 provided with the Kinect and the thermal sensor

We aim at applying a new approach to combine machine learning paradigms with computer vision techniques in order to perform image classification. Our approach is divided into three phases:

1. Computer vision transformations. The main goal of this phase is to have variability in the aspect the picture offers, so that different values are obtained for the same pixel positions. To achieve this, we combine some standard image related algorithms (edge detection, gaussian filter, binarization, and so on) in order to obtain different views of the images. Figure 2 shows an example, in which some of the transformations are used. From the original training database collected, a

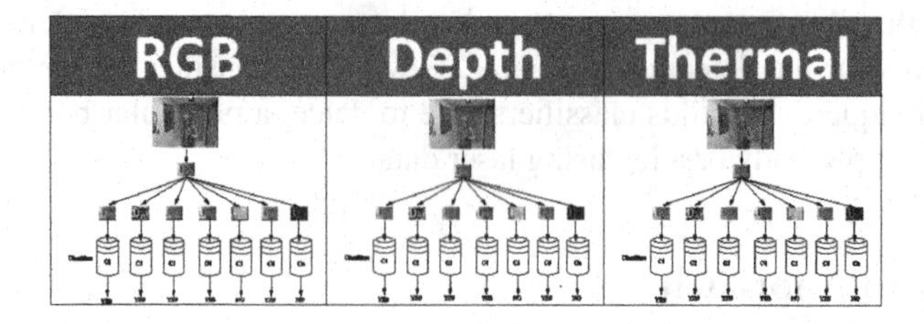

Fig. 2 Chosen approach: learning classifiers from three transformed data sources

new training database is obtained for each of the computer vision transformation used, summing up a total of 24 databases for each device.

2. In the classification phase, the system learns a classifier from a hand-labeled dataset of images (abovementioned original and transformations). As classifiers we use of five well known ML supervised classification algorithms with completely different approaches to learning and a long tradition in different classification tasks: IB1, Naive-Bayes, Bayesian Network, C4.5 and SVM.
3. Combination of classifiers, in order to finally classify the targets as human or non human. After building the individual classifiers ($5 \times 24 = 120$ for each device) the aim is at combining the output of the different classifiers to obtain a more robust final people detector.To achieve this, we use a bi-layer Stacked Generalization approach,[10], in which the decision of each of the three single classifiers is combined by means of another method. Figure 3 shows the typical approach used to perform a classification with this multiclassifier approach. It has to be noticed that the second layer classifier or function could be any function, including a simple vote approach among the used classifiers.

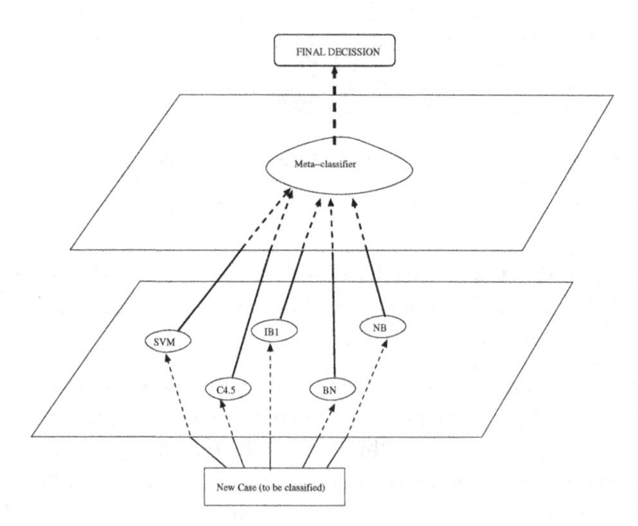

Fig. 3 Bi-layer Stacked Generalization approach

4 Experimental Setup

The manufacturing plant located at Tekniker-IK4 is a real manufacturing shop floor where machines and humans share the space in performing production activities. The shop floor can be characterized as an industrial environment, with high ceilings, fluorescent light bulbs, high windows, etc. In this context, the steps of the experimental phase are:

1. To collect a database of images that contains three data types that are captured by the two sensors. The training data set is composed of 1064 hand-labeled samples. The input to the supervised algorithms is composed of 301 positive and 763 negative examples. To obtain the positive and negative examples the robot was operated in an unconstrained indoor environment (the manufacturing plant).
2. To reduce the image sizes from 640×480 to 32×24, and convert colour images to gray-scale ones. In order to make a fast classification –real time response is expected– we first transform the colour images in gray-scale 32\times24 , and reduce as well the size of the infrared images to 32\times24 size matrix. Hence we have to deal with 768 predictor variables, instead of 307200\times(3 colours) of the original images taken by the Kinect camera.
3. For each image, to apply 23 computer vision algorithms, obtaining 23 new databases for each image type. Thus, we have 24 data sets for each image type.
4. To build 120 classifiers, applying 5 machine learning algorithms for each image type training data sets (5×24) and to apply 10 fold cross-validation using 5 different classifiers to each of the previous databases, summing up a total of $3 \times 24 \times 5 = 360$ validations.
5. To select a classifier for each image type, among its 120 different models and to make a final decision combining the results of the three classifiers selected, one for each data source.

5 Experimental Results

Performance of the people detection system is evaluated in terms of detection rates and false positives or negatives.

The table 1 shows the 10 fold cross-validation accuracy obtained using the reduced original databases. The best obtained result is 92.11% for the thermal images original database, and using SVM as classifier. The real time Kinect's algorithms accuracy among the same images was quite poor (37.50%), as the robot was moving around the environment and the Kinect has been made to be used as a static device. As a matter of fact, that has been the origin of the presented research.

The same accuracy validation process has been applied to each image transformation on each image format. The table 2 shows the 10 fold cross-validation accuracy obtained using the 24 transformed databases (only the best transformations for each machine learning algorithm). The best obtained results are: 91.83% accuracy with C4.5 classifier using Transformation 7 (T7, Gaussian one) for RGB images; again the C4.5 classifier using Transformation 7 (T7, Gaussian one) for the depth images,

Table 1 10 fold cross-validation accuracy percentage obtained for each classifier using original images.

Data source	BN	NB	C4.5	K-NN	SVM
RGB	89.20	71.74	82.63	90.89	85.35
Depth	86.29	68.64	83.29	90.89	84.04
Thermal	89.67	86.10	87.79	91.74	**92.11**

Table 2 10 fold cross-validation accuracy percentage obtained for each classifier using transformed images.

Data source	BN	NB	C4.5	K-NN	SVM
RGB	89.77 (T17)	75.12 (T16)	**91.83** (T7)	85.92 (T7)	86.29 (T6)
Depth	86.67 (T12, T20)	78.87 (T3)	**92.86** (T7)	85.26 (T7)	85.35 (T10,T16)
Thermal	90.99 (T2)	88.08 (T9)	92.77 (T5)	91.46 (T2)	**93.52** (T8,T9)

Table 3 Best three combination: 10 fold cross-validation accuracy percentage obtained.

Three Best	BN	NB	C4.5	K-NN	SVM	Vote
Results	95.02	95.49	95.59	95.40	94.84	**95.96**

with a 92.86% accuracy; we obtain the best result (93.52%) for the SVM classifier, for two of the used transformations (T8 –Lat– and T9 –Linear-strech–) for thermal images.

The last step is to combine the results of the best three classifier obtained, one by each sensor. To do that, we use a Stacking classifier in which the decision of each of the three single classifiers is combined by means of another classifier. We have also performed a single vote process among those three decisions. Table 3 shows the obtained results. As it can be seen, the best obtained accuracy is 95.96%, using the single vote approach. It significantly improves the result of the best previous classifiers (93.52 for the Thermal images).

It can be noticed that the percentage of wrong classified is drastically reduced with the classification algorithms and with the combination of the three information sources. As the Kinect algorithm provides a 62.5% of wrong classified, the fusion only a 4,98%., best RGB 7,6%, best depth 7,14% and best thermal 6,49%.

6 Conclusions and Future Works

This paper presented a people detection system for mobile robots using using 3D camera and thermal vision and provided a thorough evaluation of its performance. The main objective of this article is to evaluate the combination of machine learning paradigms with computer vision techniques to fuse multimodal information. Experimental results carried out in a manufacturing shop-floor show that the percentage of wrong classified using only Kinect is drastically reduced with the classification algorithms and with the combination of the three information sources.

Using the Kinect sensor in the mobile platform show some limitations since Kinect algorithms rely on the detection of human activities captured by a static camera. In mobile robot applications the sensors setup is assumed to be embedded in the robot that is usually moving. We showed that the detection of a person is improved by cooperatively classifying the feature matrix computed from the input data, where we made use of computer vision transformations and supervised learning techniques to obtain the classifiers. Our algorithm performed well across a number of experiments in a real manufacturing plant. In the near future we envisage to extend this implementation toward a museum scenario where we are going to develop a people tracking behaviour combining/fusing visual cues using particle filter strategies.

Acknowledgements. The work described in this paper was partially conducted within the ktBOT project and funded by KUTXA Obra Social.

References

1. Heimann sensor, `http://www.heimannsensor.com/index.php`
2. Kinect sensor, `http://en.wikipedia.org/wiki/Kinect`
3. Bellotto, N., Hu, H.: Multisensor data fusion for joint people tracking and identification with a service robot. In: 2007 IEEE International Conference on Robotics and Biomimetics ROBIO, pp. 1494–1499 (2007)
4. Cielniak, G.: People tracking by mobile robots using thermal and colour vision. Ph.D. thesis, Department of Technology Orebro University (2007)
5. Gundimada, S., Asari, V.K., Gudur, N.: Face recognition in multi-sensor images based on a novel modular feature selection technique. Inf. Fusion 11, 124–132 (2010)
6. Hofmann, M., Kaiser, M., Aliakbarpour, H., Rigoll, G.: Fusion of multi-modal sensors in a voxel occupancy grid for tracking and behaviour analysis. In: Proc. 12th Intern. Workshop on Image Analysis for Multimedia Interactive Services (WIAMIS), Delft, The Netherlands (2011)
7. Martinez-Otzeta, J.M., Ibarguren, A., Ansuategui, A., Susperregi, L.: Laser Based People Following Behaviour in an Emergency Environment. In: Xie, M., Xiong, Y., Xiong, C., Liu, H., Hu, Z. (eds.) ICIRA 2009. LNCS, vol. 5928, pp. 33–42. Springer, Heidelberg (2009)
8. Mozos, O.M., Kurazume, R., Hasegawa, T.: Multi-part people detection using 2D range data. International Journal of Social Robotics 2(1), 31–40 (2010)
9. Schiele: Visual People Detection - Different Models, Comparison and Discussion. In: IEEE International Conference on Robotics and Automation (ICRA) (2009)
10. Sierra, B., Serrano, N., Larrañaga, P., Plasencia, E.J., Inza, I., Jiménez, J.J., Revuelta, P., Mora, M.L.: Using bayesian networks in the construction of a bi-level multi-classifier. A case study using intensive care unit patients data. Artificial Intelligence in Medicine 22(3), 233–248 (2001)
11. Treptow, A., Cielniak, G., Duckett, T.: Active people recognition using thermal and grey images on a mobile security robot. In: Proceedings of the 2005 IEEE/RSJ International Conference on Intelligent Robots and Systems (IROS 2005), Edmonton, Canada (2005)
12. Zin, T.T., Takahashi, H., Toriu, T., Hama, H.: Fusion of infrared and visible images for robust person detection. Image Fusion (2011), InTech `http://www.intechopen.com/articles/show/title/fusion-of-infrared-and-visible-images-for-robust-person-detection`

A Context-Based Surveillance Framework for Large Infrastructures

Oscar Ripolles, Julia Silla, Josep Pegueroles, Juan Saenz, José Simó, Cristina Sandoval, Mario Viktorov, and Ana Gomez

Abstract. In this paper we present the control and surveillance platform that is currently being developed within the ViCoMo project. This project is aimed at developing a context modeling system which can reconstruct the events that happen in a large infrastructure. The data is presented through a 3D visualization where all the information collected from the different cameras can be displayed at the same time. The 3D environment has been modeled with high accuracy to assure a correct simulation of the scenario. A special emphasis has been put on the development of a fast and secure network to manage the data that is generated. We present some initial results obtained from seven cameras located in a Port Terminal in Spain.

1 Introduction

Computer vision is a research area which has been investigated for a long time. As a result, it is possible to find in the literature and also in the market numerous

Oscar Ripolles · José Simó
Inst. of Control Systems and Industrial Computing (ai2),
Universitat Politécnica de Valencia, Spain
e-mail: oripolles@ai2.upv.es, jsimo@disca.upv.es

Julia Silla · Cristina Sandoval
Visual-Tools, Spain
e-mail: mjsilla@visual-tools.com, csandoval@visual-tools.com

Josep Pegueroles · Mario Viktorov
Dep. d'Enginyeria Telemàtica, Universitat Politècnica de Catalunya, Spain
e-mail: josep.pegueroles@upc.edu,
mario.viktorov.mechoulam@gmail.com

Juan Saenz
Acciona Trasmediterránea, Spain
e-mail: jsaenz@acciona.es

Ana Gomez
Acciona Infraestructuras, Spain
e-mail: agomez40@acciona.es

P. Novais et al. (Eds.): Ambient Intelligence - Software and Applications, AISC 153, pp. 35–42.
springerlink.com

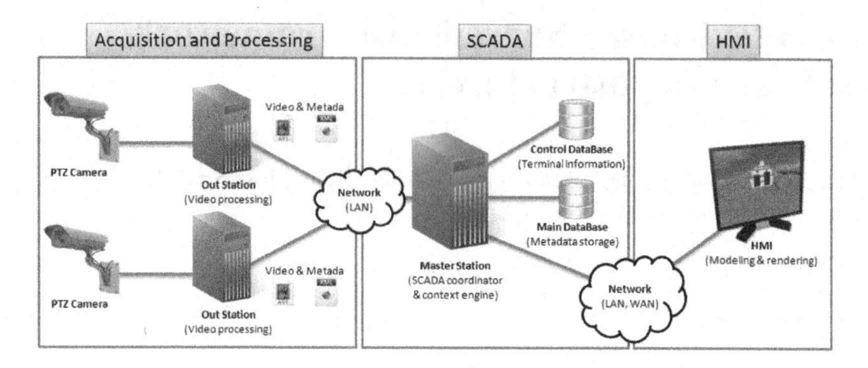

Fig. 1 Main workflow of the ViCoMo proposed platform.

computer-vision-based systems that relieve humans from a cumbersome work and also improve efficiency, security and safety. Consequently, it would be desirable to develop intelligent surveillance systems that supported this kind of tasks. A visual context modeling system could be used to construct realistic context models to improve the decision making.This context would be a perfect candidate to be displayed within a 3D reconstruction of the scenario to be analyzed. The proposed approach can be useful, for example, to follow a moving object or to select an appropriate camera position to visualize the 3D scenario from the most adequate point of view.

There have been some previous attempts to develop such a system, where 3D information is used to improve surveillance. On the one hand, some authors propose combining 3D environments with real videos. These solutions aim at offering situational awareness so that the user has a clearer understanding of where the cameras are located [16, 14] or to offer an easy navigation between adjacent cameras [7]. On the other hand, some authors propose using computer graphics techniques to estimate the position of the objects. Authors commonly use very rough 3D scenarios and create a simple textured polygon to represent the object using images from the video source [8, 14, 15].

In this paper we present the ongoing results of a research project which has the challenging aim of offering visual interpretation and reasoning using context information. The research project[1] is developing a global context modeling system with the final purpose of finding the context of events that were captured by cameras or image sensors. Then, the extracted context is modeled so that reliable reasoning about an event can be established. Finally, modeling these events and their surroundings in a very detailed 3D scenario supposes an integral part of the system, offering new means for visualization. Moreover, having a 3D feedback can improve the context modeling as it would be possible to find interpretation errors and impossible situations that, without 3D information, would not be possible to detect. An important aim of this project is the development of a trial platform for logistic control and surveillance of a large infrastructure.

[1] ViCoMo stands for Visual Context Modeling (MICINN Project TSI-020400-2011-57, ITEA2 Project IP08009).

We offer the results obtained in the Port Terminal of Acciona in Valencia (Spain), which has 50,000 m^2 perimeter and 2 docks whose mooring lines are over 300 meters long. Although the proposed framework is mainly aimed at large infrastructures, it could be scaled and used in a more reduced space to give way to smart homes.

This paper is structured as follows. In Section 2 we offer an overview of the whole ViCoMo platform. Section 3 presents a detailed description of the main components of the platform. Finally, in Section 4 we conclude on the solution proposed and outline the main lines for future work.

2 Details of the ViCoMo Platform

To achieve these purposes, the ViCoMo system is comprised of multiple cameras, a communication network, a context engine that creates the context model, a database, and a client for retrieval and navigation through the content. The visualization client must show present events in real-time as well as past events on user's demand. This application offers an augmented virtuality environment where the 3D simulation is combined with real-life information and images.

Figure 1 offers a diagram that covers the main elements of the proposed platform. In the scenario we propose, we call *Out Stations* to the computers that receive and process video streams. The *Out Stations*, which can be embedded in the cameras themselves, send periodically the output of the processing algorithms (usually as metadata) in response to the requests performed by a computer (the *Master Station*). Once the information arrives at the *Master Station*, it is stored in a historical database in order to be able to consult it afterwards. Additionally, and depending on the system configuration, the video streams and metadata can be relayed to one or several machines (labeled as *HMI* in Figure 1) for video surveillance purposes. In this Section we describe these three main stages of this platform.

2.1 Acquisition and Processing

The first stage of the system consists in analyzing the video images to extract metadata. The analysis involves three main tasks:

- **Tracking.** We have developed a multi-camera tracking algorithm that is based on a combination of motion detection and background subtraction techniques [18]. Nevertheless, the high density of vehicles during the load/unload operation of a ship makes the standard tracking algorithms fail due to occlusions. This is the reason why we propose a three level processing using tracking information along time:
    1. Intra-camera: for each track of the objects we remove impossible behaviors, such as small splits coming from a unique object and ending in a unique object. We also remove objects of small duration or with impossible shapes [3].
    2. Multi-camera: we use the method described in [6] to pair objects between views in a multi-camera system with overlapped fields of view. After that, we

apply the same criteria used for improving intra-camera information. In this case, if an object is paired with an object in another view but, at one instant, it is paired with two objects, we can fix the error by removing the wrong pairing.
3. 3D model: there is a final processing in the 3D representation that is capable of correcting the tracking results using the information of the 3D environment (e.g. a car cannot be over the water, a ship is always over the water, etc.).

- **License Plate Recognition (LPR).** LPR is an interesting feature that enables the system to link a detected object with information provided by the Port Terminal, like driver's id, destination, type of cargo, responsible company, etc. The basis of the LPR for the vehicles in the Port Terminal is the open source Optical Character Recognition (OCR) engine Tesseract [9, 17], currently developed by Google. It is considered as one of the most accurate free OCR engines available and it is able to process many different languages. In the case of plate reading, there are specific syntaxes that can be present. For doing that, Tesseract can be adapted to be able to recognize the characters that follow the desired formats.
- **Classification.** In some cases the information regarding the type of object is necessary. Together with the LPR data, this information is useful for obtaining some statistics about the presence and number of operations of a vehicle. Using basic parameters that can be extracted from the objects, such as the shape, area and position, we are able to classify between truck, vehicle, person and group of people. For the classification we have tried different classifiers from the Machine Learning libraries from OpenCV [1] and the best classification results for our sequences are achieved by the random tree classifier.

2.2 *SCADA and Data Management*

To achieve the data transmission and storage in a scalable fashion a Supervisory Control and Data Acquisition (SCADA) system is used. SCADAs have been extensively used for data mining, control and management of resources or monitoring and surveillance. However, the SCADA system proposed in ViCoMo innovates in the way information security is supported.

As SCADA protocol we chose the DNP3 [11], which is free and evolving into an increasingly complex protocol with more features. Another important decision to make is deciding how cameras are handled from the SCADA. Usually each manufacturer provides the camera with proprietary software and protocols to manage them. Nevertheless, IP cameras manufacturers are divided into two large consortia to create standard protocols: ONVIF [12] and PSIA [4]. Currently, there is a competition to see who specifies the best standard and win the battle. Since the end of the battle is not clear we decided to integrate both (ONVIF and PSIA) in the ViCoMo project. Finally, we complete the solution with the possibility of sending video streams using protocols designed for this purpose (RTP).

One of the cornerstones of the work developed involves integrating the networking protocols with the output of the video processing protocols. The interfaces between the different components of the proposed platform (see Figure 1) have been implemented as follows:

- between the cameras and the *Out Stations*: it is done through Ethernet cabling, using protocols such as MPEG4 or JPG over HTTP or RTP.
- between the *Out Stations* and the *Master Station*: it is done using DNP3 including the result of video processing in specific containers deposited within the *Out Station* in XML format. These files are parsed and embedded in DNP3 frames.
- between *Master Station* and the database: the data must be extracted from the DNP3 frame and entered to the database using a single channel whose session remains open to facilitate operations.
- between the *Master Station* and the *HMIs*: it is done via DNP3, analogously, on the other end data is extracted and passed to the TCP layer - an entry point to the libraries of the 3D rendering software.

Given the relevance of the data transmitted in ViCoMo, it is necessary to ensure encrypted communication. We originally thought about using security mechanisms such as high level TLS, which would encrypt the streams individually, in combination with a chain of trust or certificates for each component. However, several drawbacks raised: the need to establish encrypted connections for each connection, the computational burden for the *Master Station* to manage hundreds of encrypted flows or the delay introduced by public key cryptography.These were the reasons why we decided to secure the communication channel by encrypting all the communications with IPSec. Although having the security at a lower layer results in a small overhead in traffic, the benefits of IPSec are propagated to all other applications and protocols that are used.

2.2.1 Context Engine

One of the main contributions of the ViCoMo platform is the development of a context engine, which derives simple conclusions with the information stored in the databases and acquired from the cameras. For limiting the scope of the context extraction, different use cases have been defined so that the proposed platform is applied from a functional point of view. Use cases defined in the Port Terminal scenario come up from the analysis of opportunity of technology involved in the research project, in order to improve the exploitation of port activities through the reduction of operative costs and improved security conditions. As an example of these use cases, Figure 2 presents two possible applications. On the one hand, Figure 2(a) shows an example of tracking vehicles; on the other hand, in Figure 2(b) a risk detection example is shown where a person without safety vest is traversing the terminal. In the Port Terminal we are going to detect different types of events:

- Access control: this is done by comparing the license plate of each vehicle at the entrance of the terminal with a database of unauthorized vehicles.
- Cargo loading and unloading operations: using the tracking and classification information it is possible to monitor the number of load and unload operations of

(a) Cargo and vehicles tracking.

(b) Risk situation detection.

Fig. 2 Example of use cases covered by the proposed platform.

a ship. We do this by using the tracking information and defining a virtual barrier to count the number of objects in each direction.

- Passenger boarding and landing monitoring: we use the method described in [2] for counting the number of people boarding and landing. The results are compared with the database of passengers that boarded in the previous port to check that everybody has left the ship.
- Risk situation detection: the tracking information allows us to detect some risky situations, such as the detection of people or vehicles in forbidden areas.

2.3 HMI and 3D Modeling

The visualization system that ViCoMo proposes must display 3D scenes enriched with the context information mentioned in the previous sections. This way, the rendering engine must be capable of displaying non 3D information like text of live video streams acquired from the cameras. The HMI we propose is based on the OpenSceneGraph (OSG) rendering engine [13], which includes some interesting features for visualizing 3D information. Taking into consideration that the ViCoMo project requires very precise 3D models, we decided to use Building Information Modeling (BIM) software. All objects/models have been based on CAD plans which must be precise in their measurements. In our case we used ArchiCAD [10], and additional texturing was made with 3DStudio [5] using real pictures taken from the Terminal.

The HMI application must be capable of retrieving information from the SCADA. The data repository presents an interface which offers different queries. Firstly, we

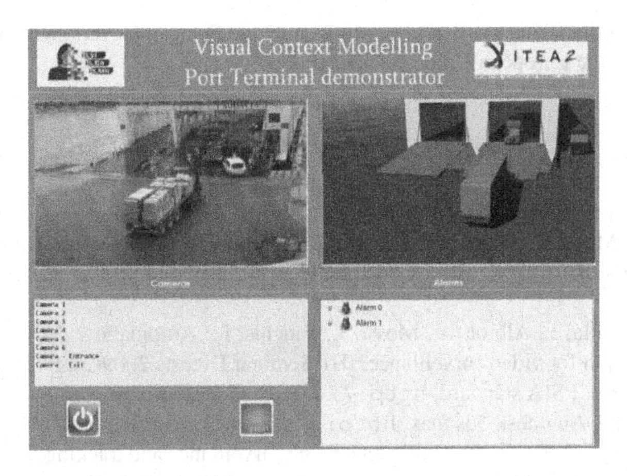

Fig. 3 HMI proposal for ViCoMo.

must be able to know, for a given timestamp, which elements are present at the scenario. This query is necessary, for example, when initializing the simulation or when accessing historical data. Then, the application must also be able to update the elements on the scenario. This second query uses a time interval (defined by two timestamps) to offer the required information. Figure 3 presents a proposal for HMI where the 3D scene reconstruction from a real image can be observed, displaying the information from a snapshot where three trucks are detected.

3 Conclusions

In this paper we have presented the ongoing work of the ViCoMo research project. This project aims at developing a new control and surveillance platform which models a context of the events to allow for simple conclusions. Moreover, an improved HMI application has been described, as it will still be necessary to have human interaction for managing logistics control and surveillance. The user application triggers alarms to get the attention from the surveillance staff to act when a set of events are detected. When managing these alarms, the system can offer a 3D simulation of the environment as well as real time camera access.

There have been some previous attempts to develop similar systems, but our proposal offers a complete solution for large infrastructures using an accurate 3D visualization system. As future work we would like to continue developing this platform, making a special effort to exploit the 3D context to improve the results from the image-processing algorithms. Moreover, the reasoning unit is continuously updated with new rules and actors to improve the context analysis and extend the functionalities of the platform.

Acknowledgements. This work has been funded by the Spanish Government (TSI-020400-2011-57) and by the European Union (ITEA2 IP08009).

References

1. Open Source Computer Vision, http://opencv.willowgarage.com/wiki
2. Albiol, A., Mora, I., Naranjo, V.: Real-time high density people counter using morphological tools. IEEE Transactions on Intelligent Transportation Systems 2(4), 204–217 (2001)
3. Albiol, A., Silla, J., Albiol, A., Mossi, J., Sanchis, L.: Automatic video annotation and event detection for video surveillance. IET Seminar Digests 2009(2), P42 (2009)
4. P.S.I. Alliance. PSIA standard, http://www.psialliance.org
5. Autodesk Inc. Autodesk 3ds max, http://usa.autodesk.com/adsk
6. Black, J., Ellis, T., Rosin, P.: Multi view image surveillance and tracking. In: Proceedings of the Workshop on Motion and Video Computing, MOTION 2002, p. 169 (2002)
7. de Haan, G., Scheuer, J., de Vries, R., Post, F.: Egocentric navigation for video surveillance in 3D virtual environments. In: IEEE Workshop on 3D User Interfaces, pp. 103–110 (2009)
8. Fleck, S., Busch, F., Biber, P., Strasser, W.: 3D surveillance a distributed network of smart cameras for real-time tracking and its visualization in 3D. In: CVPRW 2006, p. 118 (2006)
9. Google Inc., Tesseract OCR, http://code.google.com/p/tesseract-ocr
10. Graphisoft. ArchiCAD 15, http://www.graphisoft.com/products/archicad
11. D.U. Group. Distributed Network Protocol (DNP3), http://www.dnp.org
12. ONVIF. ONVIF core specification ver 2.1. (2011), http://www.onvif.org
13. Osfield, R., Burns, D.: OpenSceneGraph, http://www.openscenegraph.org
14. Rieffel, E.G., Girgensohn, A., Kimber, D., Chen, T., Liu, Q.: Geometric tools for multicamera surveillance systems. In: IEEE Int. Conf. on Distributed Smart Cameras (2007)
15. Sebe, I., Hu, J., You, S., Neumann, U.: 3D video surveillance with augmented virtual environments. In: ACM SIGMM Workshop on Video Surveillance, pp. 107–112 (2003)
16. Sentinel AVE LLC. AVE video fusion (2010), http://www.sentinelAVE.com
17. Smith, R.: An Overview of the Tesseract OCR Engine. In: Proceedings of the 9th International Conference on Document Analysis and Recognition, vol. 2, pp. 629–633 (2007)
18. Yilmaz, A., Javed, O., Shah, M.: Object tracking: A survey. ACM Comput. Surv. 38, 1–45 (2006)

Orientation System for People with Cognitive Disabilities

João Ramos, Ricardo Anacleto, Ângelo Costa, Paulo Novais,
Lino Figueiredo, and Ana Almeida

Abstract. In health care there has been a growing interest and investment in new tools to have a constant monitoring of patients. The increasing of average life expectation and, consequently, the costs in health care due to elderly population are the motivation for this investment. However, health monitoring is not only important to elderly people, it can be also applied to people with cognitive disabilities. In this article we present some systems, which try to support these persons on doing their day-to-day activities and how it can improve their life quality. Also, we present an idea to a project that tries to help the persons with cognitive disabilities by providing assistance in geo-guidance and keep their caregivers aware of their location.

Keywords: Cognitive disabilities, mobile communication, localization, persons tracking, ambient intelligence.

1 Introduction

Due to technological advancements the average life expectancy has been increased. The number of elderly people is increasing thus, the populations are getting older. The elderly population requires, among others, more health care, which in some cases involves the existence of a caregiver (family or not). If this caregiver cannot assist the elderly person at home, then the elderly person is forced to move to a family member's or nursing home.

João Ramos · Ângelo Costa · Paulo Novais
Informatics Department, University of Minho, Portugal
e-mail: a52547@alunos.uminho.pt, acosta@di.uminho.pt,
pjon@di.uminho.pt

Ricardo Anacleto · Lino Figueiredo · Ana Almeida
GECAD - Knowledge Engineering and Decision Support, ISEP, Portugal
e-mail: rmao@isep.ipp.pt, lbf@isep.ipp.pt, amn@isep.ipp.pt

P. Novais et al. (Eds.): Ambient Intelligence - Software and Applications, AISC 153, pp. 43–50.
springerlink.com

Any of the options described above involves loss of independence by the person who requires health care (elderly). In order to maintain an independent life, smart homes are considered a good alternative [13]. Besides elderly, smart houses can be also used by people with cognitive disabilities. Several intelligent devices embedded in home environment can provide assistance to the resident, monitoring his movements and his health 24 hours a day [15, 4].

Smart houses besides being physically versatile are also user friendly, *i.e.*, perform their functions without bother, be inconvenient or restrict user movements [4]. Its goal is therefore to provide comfort and pleasure to the person.

Smart house's are developed based on the concept of Ambient Intelligence (AmI). This concept was firstly introduced by Information Society Technologies Advisory Group (ISTAG) [11]. In a simple way, AmI can be defined as the combination of ubiquitous computing with adaptive user interfaces [8]. The goal of AmI is to develop sensitive environments that are responsive to the presence of humans. To achieve this goal it is necessary to integrate these environments with ubiquitous computing [13].

Since 1988 the interest and the attention on how assisted technologies can improve the functional needs of people with cognitive disabilities has grown. The development of several projects related to this technology has increased the level of interest and the public awareness to the usage of assisted technology by people with cognitive disabilities and how technology can improve their lives [2].

Like an ordinary person, individuals with cognitive disabilities leave their homes, but once outside, the security and automated actions of their smart houses are inviable. In the last years, the interest and technology improvement has turned people's attention to assisted technology outside home and not only inside.

The integration of technology in day-to-day human life allows an interaction between the system and the person who uses it. In order to totally achieve the concept of AmI, it is necessary to turn the implicit communication (that prevails today) into explicit. In this way, concepts such as ubiquitous computing and invisible computing arise [14, 7]. The interaction should no longer be a simple use of interfaces, becoming more interactive, recognizing humans presence, adaptive and responding to the needs, habits, gestures or emotions shown by the residents.

Nowadays, AmI is integrated in smart houses, which contains several automated devices that may be remotely controlled by the user. Their main goal is to provide comfort, energy saving and security to the residents. This can be achieved by domestic tasks automatization [15].

New developed devices have to be easy to use, lightweight, small and resistant. If these conditions are not verified there is a high probability that the device will not be used for a long time. The person with cognitive disabilities will lose the interest and, therefore, will not use it.

In section 2 we present in detail some of the work developed in this area. In section 3 we describe our propose for a system that tries to help people with cognitive disabilities. Finally, at section 4 we make a brief conclusion of this paper.

2 State of the Art

Dawe, in [9, 10], presents how people with cognitive disabilities interact and communicate using a cell phone. They keep in touch with other persons, family and caregiver with this device. Dawe shows that many of assisted devices are acquired, but are not successfully adapted to the needs of people with cognitive disabilities. This failure may come from complex utilization and handling.

The developed application has simple interfaces loaded with pictures that act like buttons to perform or receive calls, and has a voice memo which alerts the mobile user for an incoming activity. It was also developed a time-based reminder that is activated at a certain time. This reminder is based on daily calls that parents use to remind the person with cognitive disabilities what he had to do next.

Dawe has stimulated the participation of cognitive disability people in the application development, where it was concluded that the system is useful to improve their capability to interact with other persons.

Carmien *et al.* [5] developed a system to support people with cognitive disabilities to travel using public transportation. They studied several techniques that caregivers and people with cognitive disabilities use to travel, for example, from home to school. The simplest way is using a standard bus map and then create a personal map with the important routes colored (figure 1 a)). In addition to routes, significant landmarks are also marked on the map. Moreover, each person with the cognitive disability has a group of time cards (figure 1 b)). These cards, according to the authors, are used if that person wants to go, for example, to the shopping center. The person with cognitive disabilities finds the correct card which has information about the bus route and schedule.

After this analysis they developed a system architecture where the person with cognitive disabilities just has to use a PDA, instead of the map and time cards. This architecture has several goals, the first is to give assistance to the mobile user, sending just-in-time information for many tasks, including destination and information about the correct bus. The second is, when the user needs, start a communication

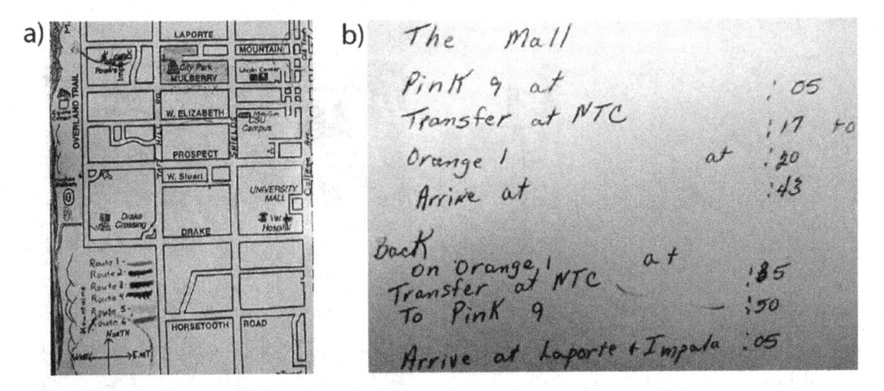

Fig. 1 Traditional artifacts personalized during training [5] - a) Personal colored map; b) Time cards

between him and the caregiver. Finally, the system supplies a safety functionality if something goes wrong.

Based on this architecture, Carmien *et al.* developed a prototype which can be seen on figure 2. On the figure left side it is presented a menu with possible destination options, while the figure on the right side shows bus position on real-time. The buses on the street are equipped with GPS and send wirelessly their position to a server. Bus agents have access to this information and update bus position on the map. When the user selects the destination the system calculates the routes based on the buses GPS information.

There are Mobility Agents that detect if the user has boarded the correct bus. In case of missing the bus or taking the wrong bus, these agents use heuristics to correct the situation.

The goal of this prototype is to give to the mobile device the capability to generate visual and audio prompts caused by real-world events. With this device people with cognitive disabilities can use the public transportation system without getting lost, or taking the wrong bus.

The Personal Travel Assistant prototype was based on mobility agents. These agents are those who really help people with cognitive disabilities to travel using public transportation system. Although traveling from one place to another is an important task in people's life, this usually represents a sub-activity of their day. To implement a broader prompting system the authors developed MAPS - Memory Aiding Prompting System.

The creation of scripts by the caregivers is possible due to an end-user programming tool. With this tool caregivers can create several scripts according to the activities that mobile user wants or needs to do. These scripts have audio-visual stimulation to keep the attention of the mobile user (person with cognitive disabilities).

Recently, Chu *et al.* [6] developed an assistance system that has an interactive activity recognition and prompting. The model has a learning process that adapts the prompting system to each user, helping in their daily activities. The schedule is then, if necessary, revised as many times as the user needs.

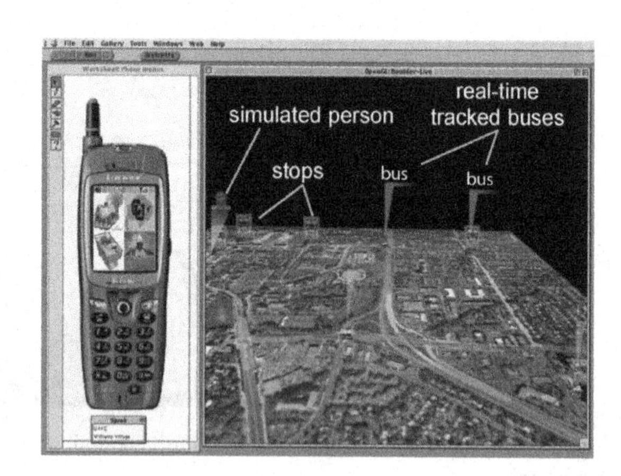

Fig. 2 Prototype showing a prompting device (on left) and a real-time bus system (on right)[5]

Fig. 3 Proposed System Demonstration

The developed system use data from sensors (*e.g.*, IR motion sensors, RFID touch sensors) to determine the user's state. When the system is uncertain of the user state then it queries the user to know what he or she is doing. In order to avoid constant interruptions from questions or prompts, Chu *et al.* proposed a selective-inquiry rule. This interrupts the user only when trying to reduce the uncertainty.

3 System Description

In the previous sections we have shown that a mobile device (*e.g.*, cell phone or smartphone) can be very useful in the day life of people with cognitive disabilities. For instance, caregivers use it to communicate and to know the current location of the person who they are taking care .

We propose a system that tries to support people with cognitive disabilities and to keep their caregivers more unconcerned.

This project consists on an orientation system to help people with cognitive disabilities, turning them more independent. This orientation system consists on a mobile application, which presents, along the user travel path, several landmarks (pictures) to help him to know which direction he must take next, as we demonstrate on figure 3. In the case that a picture is not available, then an arrow will appear indicating the correct path. During an activity or a travel the user will receive, if necessary, prompts indicating several types of information. This information is given just-in-time and includes directions (*e.g.*, turn left) or reminders to keep the user's attention in the present action.

These directions are given by the localization system, which retrieves the person's location and verifies if the person is performing the correct travel path.

If, for some reason, there is an emergency or the user gets lost then he only has to press a simple panic button in the smartphone home screen. When activated it initiates a call and/or sends a SMS to the caregiver with the current user location.

We will develop a web and another Android application to the caregiver be aware, in real-time, of the actions of the person with cognitive disabilities (*e.g.*, where he is going or what is doing).

This system will be developed for Android Operating System and it will use mobile device GPS module to retrieve user outdoor location. However, indoors or in a more dense environment (big cities with tall buildings, dense forests, etc) GPS does not work or does not provide satisfactory accuracy. So, we have a problem, in indoor environments the proposal of only use GPS to retrieve location, does not work. To suppress this limitation we will integrate GPS with an Inertial Navigation System (INS) presented on [3]. Using this type of localization system the user is not context dependent (*e.g.*, smart homes - like we have seen earlier).

The main goal of this localization system is to retrieve location in indoor environments without using a structured environment and in a non-intrusive way. It consists on several sensors (Body Sensors Units) integrated on person's clothes and shoes, spread over the person's lower limbs. These units are connected to a Body Central Unit (BCU), which receives the sensors data and handles the location estimation, to then send it to the user's mobile device. One problem is that INS can have a big drift, due to sensor's thermal changes, which leaves to error accumulations and to poor location estimations. However, we are implementing a probabilistic algorithm that learns the person walking behaviors to, in real-time, correct the sensors data.

In figure 4, we present the framework for our system. People with cognitive disabilities must have a smartphone with Android Operating System and a GPS module. Also, they can have the INS that we have presented before, which connects with the smartphone through a Bluetooth connection, giving to it the actual user location (both indoor and outdoor). This INS includes some sensors, like accelerometers, gyroscopes, force and pressure sensors. With user current location, the smartphone will process the necessary data to present, to the user, the correct landmark for his location in order to assist user's orientation.

Besides showing this information, the mobile application sends (every 5 minutes or 200 meters), over the Internet, the current user location to the system server. Thus, the server must be able to communicate in a standard way, supporting SOAP [1] and OAUTH [12]. It will have a database and decision engines that, for instance, send a message to the caregiver if the sensor platform detects an user emergency.

Finally, the caregivers can connect to the server, through the web or the mobile application to retrieve the person with cognitive disabilities current location. Also, an agent platform, that work with general web information, is used to obtain a landmark image that is needed and is not already stored in the smartphone.

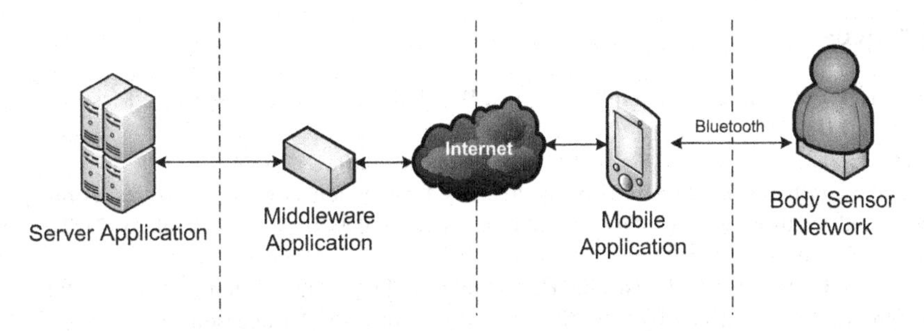

Fig. 4 Framework

4 Conclusion

Nowadays assisted technology is being increasingly used to improve the life of people with cognitive disabilities. There are several studies that analyze these people needs, presenting what is done and what needs to be done to improve their life quality.

There are several systems that assist the life of the residents of smart houses, but once outside the security and automated actions of the smart houses are not available. This is a problem to people with cognitive disabilities because they cannot be completely independent and need a caregiver.

Some authors tried to solve this issue by developing many types of applications that help people with cognitive disabilities, for example, to assist people using public transportation system, but traveling is not the only task that these people perform. It represents a sub-task (sub-activity) that needs to be accomplished in order to do an entire activity. For this reason it was developed a prompting system to assist people with cognitive disabilities and turn them more independent.

In this paper we have presented a system that tries to make these people even more independent and keep caregivers more unconcerned. The goal of our system is to provide an orientation system that works both indoor and outdoor in a non-structured environment. By the other hand we try to develop a monitoring system that gives to the caregiver, in real-time, the current location of the person with cognitive disabilities.

With this system and an INS module people with cognitive disabilities location can be known even when the GPS of the smart phone is not available. This system allows caregivers to always know the current location of the person with cognitive disabilities. This location is updated on the server in a pre-established interval of time. An agent platform that works on mobile devices is used to send a message when the agent detects an user emergency.

The system is currently in a development stage and we expect to do real tests with people in the near future.

References

1. Adams, P., Easton, P., Johnson, E., Merrick, R., Phillips, M.: SOAP over Java Message Service 1.0 (2011), `http://www.w3.org/TR/2011/PR-soapjms-20111208/`
2. Alper, S., Raharinirina, S.: Assistive technology for individuals with disabilities: A review and synthesis of the literature. Journal of Special Education Technology 21(2), 47–64 (2006)
3. Anacleto, R., Figueiredo, L., Novais, P., Almeida, A.: Providing location everywhere. In: Progress in Artificial Intelligence - Proceedings of the 15th Portuguese Conference on Artificial Intelligence - EPIA 2011, pp. 15–28 (2011)
4. Augusto, J., Mccullagh, P.: Ambient intelligence: Concepts and applications. Computer Science and Information Systems 4(1), 1–27 (2007)
5. Carmien, S., Dawe, M., Fischer, G., Gorman, A., Kintsch, A., Sullivan, J.F.: Socio-technical environments supporting people with cognitive disabilities using public transportation. ACM Transactions on Computer-Human Interaction 12(2), 233–262 (2005)
6. Chu, Y., Song, Y.C., Kautz, H., Levinson, R.: When did you start doing that thing that you do? Interactive activity recognition and prompting. In: AAAI 2011 Workshop on Artificial Intelligence and Smarter Living: The Conquest of Complexity, p. 7 (2011)
7. Costa, R., Costa, A., Lima, L., Neves, J., Gomes, P.E., Marques, A., Novais, P.: Ambient Intelligence and Future Trends. In: International Symposium on Ambient Intelligence (ISAmI 2010), vol. 72, pp. 185–188. Springer, Heidelberg (2010)
8. Costa, R., Neves, J., Novais, P., Machado, J., Lima, L., Alberto, C.: Intelligent mixed reality for the creation of ambient assisted living. In: Proceedings of Progress in Artificial Intelligence, pp. 323–331 (2007)
9. Dawe, M.: Desperately seeking simplicity: how young adults with cognitive disabilities and their families adopt assistive technologies. In: Proceedings of the SIGCHI Conference on Human Factors in Computing Systems, CHI 2006, pp. 1143–1152. ACM (2006)
10. Dawe, M.: Let me show you what I want: engaging individuals with cognitive disabilities and their families in design. In: CHI 2007 Extended Abstracts on Human Factors in Computing Systems, CHI EA 2007, pp. 2177–2182. ACM (2007)
11. Ramos, C.: Ambient intelligence - a state of the art from artificial intelligence perspective. In: Proceedings of Progress in Artificial Intelligence, pp. 285–295 (2007)
12. Recordon, D., Hardt, D., Hammer-Lahav, E.: The oauth 2.0 authorization protocol. Network Working Group, pp. 1–47 (2011)
13. Sadri, F.: Multi-Agent Ambient Intelligence for Elderly Care and Assistance. AIP, pp. 117–120 (2007)
14. Schmidt, A.: Interactive Context-Aware Systems Interacting with Ambient Intelligence, vol. 6, pp. 159–178. IOS Press (2005)
15. Stefanov, D.H., Bien, Z., Bang, W.C.: The smart house for older persons and persons with physical disabilities: Structure, technology arrangements, and perspectives. IEEE Transactions on Neural Systems and Rehabilitation Engineering 12(2), 228–250 (2004)

Proximity Detection Prototype Adapted to a Work Environment

Gabriel Villarrubia, Alejandro Sánchez, Ignasi Barri, Edgar Rubión, Alicia Fernández, Carlos Rebate, José A. Cabo, Teresa Álamos, Jesús Sanz, Joaquín Seco, Carolina Zato, Javier Bajo, Sara Rodríguez, and Juan M. Corchado

Abstract. This article presents a proximity detection prototype that uses ZigBee technology. The prototype is primarily oriented to proximity detection within an office environment and some of the particular characteristics specific to such an environment, including the integration of people with disabilities into the workplace. This allows the system to define and manage the different profiles of people with disabilities, facilitating their job assimilation by automatically switching on or off the computer upon detecting the user's presence, or initiating a procedure that automatically adapts the computer to the personal needs of the user.

Keywords: Zigbee, proximity detection, wake on LAN, personalization.

Alejandro Sánchez · Javier Bajo
Universidad Pontificia de Salamanca, Salamanca, Spain
e-mail: {asanchezyu,jbajope}@usal.es

Gabriel Villarrubia · Carolina Zato · Sara Rodríguez · Juan M. Corchado
Departamento Informática y Automática, Universidad de Salamanca, Salamanca, Spain
e-mail: {gvg,carol_zato,srg,corchado}@usal.es

Ignasi Barri · Edgar Rubión · Alicia Fernández · Carlos Rebate
Indra, Spain
e-mail: {ibarriv,erubion,afernandezde,crebate}@indra.es

José A. Cabo · Teresa Álamos
Wellness Telecom, Spain
e-mail: {talamos,jacabo}@wtelecom.es

Jesús Sanz · Joaquín Seco
CSA, Spain
e-mail: {jesus.sanz,joaquin.seco}@csa.es

P. Novais et al. (Eds.): Ambient Intelligence - Software and Applications, AISC 153, pp. 51–58.
springerlink.com

1 Introduction

Modern societies are characterized as much by technological advances as they are by social advances. With regards to technological advances, the rapid development of TIC has had a direct impact on our lifestyle, while the realm of social advances includes an increasing sensitivity to allow persons with disabilities, or those at risk of social exclusion, to lead a normal and independent life, which includes the very relevant ability to carry out and receive compensation for professional activity. The effective integration into the workplace of people with disabilities is an enormous challenge for society, and new technologies should provide solutions that can further this integration.

There are a number of obstacles that impede the ability of people with disabilities to integrate in the workforce, and the ability of businesses to include them in their staff. One of the biggest challenges of incorporating people with disabilities into the workforce is the question of autonomy and mobility.

Within the field of technologies specifically developed to facilitate the lives of people with disabilities, there have been many recent advances that have notably improved their ability to perform daily and work-related tasks, regardless of the type and severity of the disability. Nevertheless, the complete integration of these individuals in the society in general, and the workforce in particular, is still considered a challenge. Consequently, it is critical to provide new tools that can eliminate these barriers and facilitate the incorporation of this group of individuals into the workforce. This article presents a proximity detection prototype, specifically developed for a work environment, which can facilitate tasks such as activating and personalizing the work environment; these apparently simple tasks are in reality extremely complicated for some people with disabilities.

The rest of the paper is structured as follows: The next section introduces the technology used in the development of this prototype. Section 3 presents the most important characteristics of the prototype. Finally, section 4 explains some results and conclusions that were obtained.

2 Technology Applied

There are currently different types of sensors that can distinguish whether a person or object is located within its range of action. To begin, there are presence sensors whose purpose is to determine the presence of an object at a specified interval of distance. There are also proximity sensors, which can measure the distance from an initial point (usually the sensor itself) to the object within a range of action. Other types of sensors include ZigBee sensors, which have been successfully applied in schools, hospitals, homes, etc. ZigBee is a low cost, low power consumption, two-way wireless communication standard that was developed by the ZigBee Alliance [5]. It is based on the IEEE 802.15.4 protocol, and operates on the ISM (Industrial, Scientific and Medical) band at 868/915MHz and a 2.4GHz spectrum. Due to this frequency of operation among devices, it is possible to transfer materials used in residential or office buildings while only minimally affecting system performance [1]. Although this system can operate at the same frequency as Wi-Fi devices, the possibility that it will be affected by their presence is practically null, even in very noise environments (electromagnetic interference). ZigBee is designed to be embedded in consumer electronics, home and building automation, industrial controls, PC peripherals, medical sensor applications, toys and games, and is intended for home, building and industrial automation purposes, addressing the needs of monitoring, control and sensory network applications [5]. ZigBee allows star, tree or mesh topologies. Devices can be configured to act as network coordinator (control all devices), router/repeater (send/receive/resend data to/from coordinator or end devices), and end device (send/receive data to/from coordinator) [6].

The IEEE 802.15.4 standard is designed to work with low rate networks with limited resources [2]. Furthermore, ZigBee incorporates an additional network with security and application layers under the IEEE 802.15.4 standard, and allows more than 65,000 connected nodes.

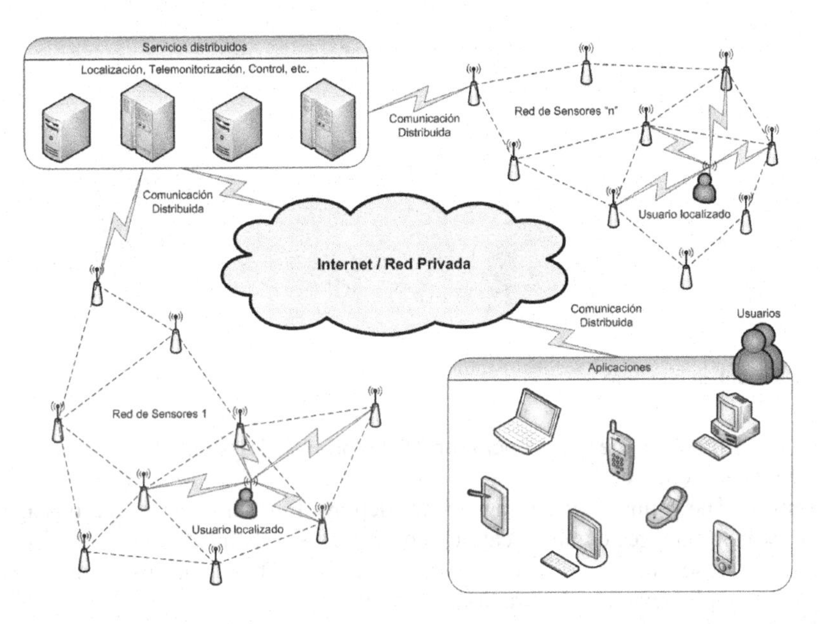

Fig. 1 Diagram of the Zigbee network

One of the main advantages of this system is that, as opposed to GPS type systems, it is capable of functioning both inside and out with the same infrastructure, which can quickly and easily adapt to practically any applied environment. The infrastructure of the proposed system is dynamic, adaptable, scalable, non-intrusive, and has a low operation cost.

3 Prototype

3.1 Architecture

This document presents a proximity detection system that will be used by people with disabilities to facilitate their integration in the workplace. The main goal of the system is to detect the proximity of a person to a computer using ZigBee technology. This allows an individual to be identified, and for different actions to be performed on the computer, thus facilitating workplace integration: automatic switch on/off of the computer, identifying user profile, launching applications, and adapting the job to the specific needs of the user.

Every user in the proposed system carries a Zigbee tag, while a ZigBee reader is located in each system terminal. Thus, when a user tag is sufficiently close to a specific terminal, the reader can detect the user tag and immediately send a message to the ZigBee network coordinator.

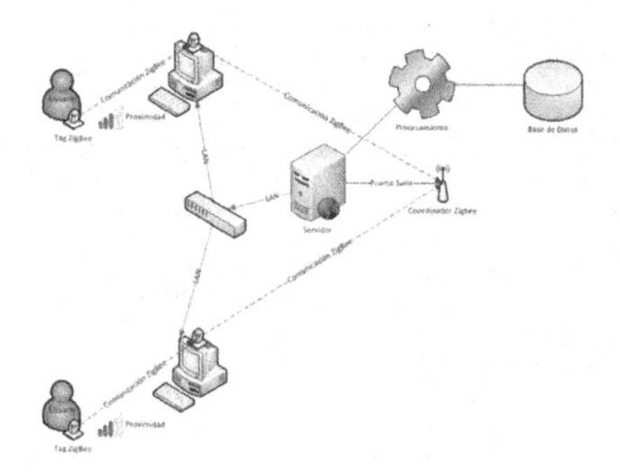

Fig. 2 System architecture

Figure 2 displays the basic operation of the proposed system. The components used in the system are:

Server: The primary system agents are deployed on this device: the manager agent, which manages communication and events; and the profile manager agent, which is responsible for managing user profiles. The agent functions were originally implemented with a traditional server.

Clients: These are user agents located in the client computer and are responsible for detecting the user with ZigBee technology, and for sending the user's identification to the manager agent. These agents are responsible for requesting the profile role adapted for the user.

Data base: The system uses a data base, which stores data related to the users, sensors, computer equipment and status, and user profiles.

ZigBee Coordinator: A ZigBee device responsible for coordinating the other ZigBee devices in the office. It is connected to the server by a serial port, and receives signals from each of the ZigBee tags in the system.

ZigBee Reader: These are ZigBee devices that are used to detect the presence of a user. Each ZigBee reader is located in a piece of office equipment (computer).

ZigBee Tag: These are ZigBee devices carried by the user. When a tag is located near a reader (within a range defined according to the strength of the signal), it initiates the process of detecting proximity.

Network infrastructure: The system uses a LAN infrastructure that uses the wake-on-LAN protocol for the remote switching on and off of equipment.

3.2 Wake On Lan

Wake-on-LAN/WAN is a technology that allows a computer to be turned on remotely by a software call. It can be implemented in both local area networks (LAN) and wide area networks (WAN) [4]. It has many uses, including turning on a Web/FTP server, remotely accessing files stored on a machine, telecommuting, and in this case, turning on a computer even when the user's computer is turned off [7].

The Wake On LAN protocol defines a package called "Magic Package". This package contains the MAC address of the machine that it is desired to switch on. The Magic Package is sent by the link data layer to all the NICs (Network Interface Controller) using the address of the network diffusion. The frame is formed by 6 bytes (FF FF FF FF FF FF) followed of 16 repetitions of 48 bits that represents the MAC address. The frame is 102 bytes in total.

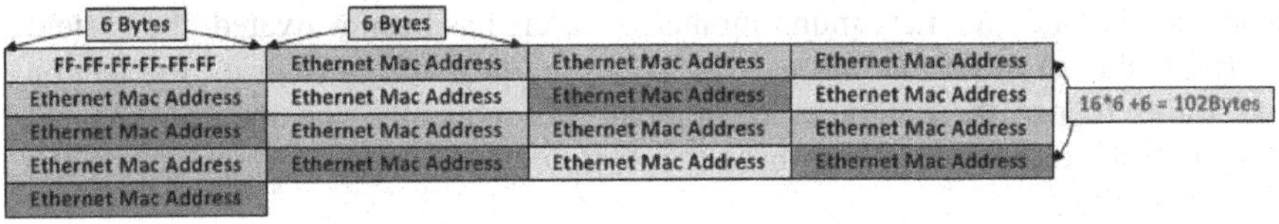

Fig. 3 Wake-on-LAN magic packet

3.3 Detection System

The proposed proximity detection system is based on the detection of presence by a localized sensor called the control point, which has a permanent and known location. Once the object has been detected and identified, its location is delimited within the proximity of the sensor that identified it. Consequently, the location is based on criteria of presence and proximity, according to the precision of the system and the number of control points displayed.

The parameter used to carry out the detection of proximity is the RSSI (Received Signal Strength Indication), a parameter that indicates the strength of the received signal. This force is normally indicated in mW or using logarithmic units (dBm). 0 dBm is equivalent to 1mW. Positive values indicate a signal strength greater than 1mW, while negative values indicate a signal strength less than 1mW.

Under normal conditions, the distance between transmitter and receiver is inversely proportional to the RRSI value measured in the receiver; in other words, the greater the distance, the lower the signal strength received. This is the most commonly used parameter among RTLS.

As shown in Figure 3, the tag located between two readers will be recognized by each reader with different RSSL levels.

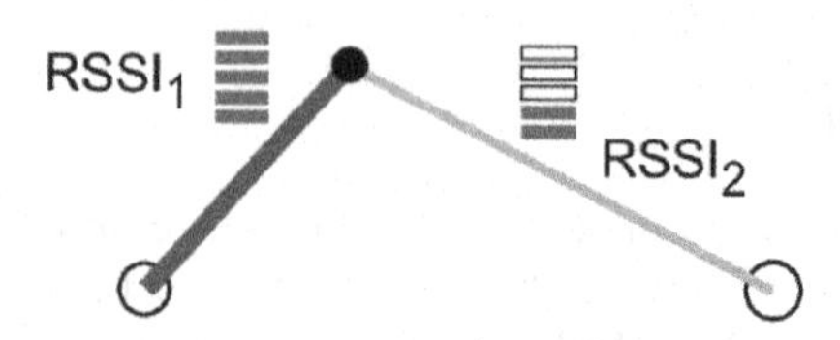

Fig. 4 Diagram of proximity distance detection based on RSSI

RSSI levels provide an appropriate parameter for allowing our system to function properly. However, variations in both the signal transmission and the environment require us to define an efficient algorithm that will allow us to carry out our proposal. This algorithm is based on the use of a steps or measurement levels (5 levels were used), so that when the user enters the range or proximity indicated by a RSSI level of -50, the levels are activated. While the values received are less than the given range, each measurement of the system activates a level. However, if the values received fall outside the range, the level is deactivated. When the maximum number of levels has been activated, the system interprets this to mean that the user is within the proximity distance of detection and wants to use the computer equipment. Consequently, the mechanisms are activated to remotely switch on both the computer and the profile specific to the user's disability.

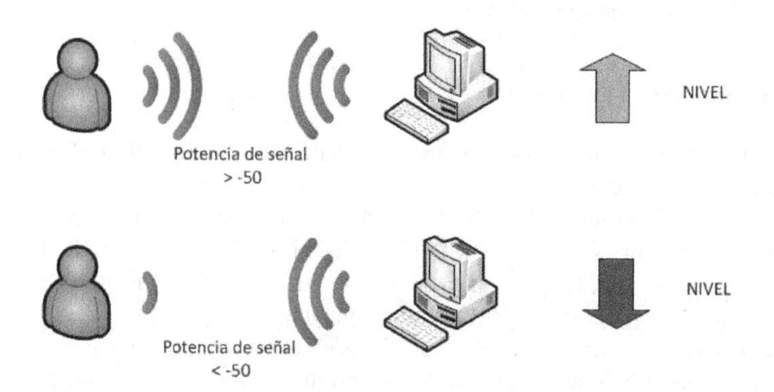

Fig. 5 System operation according to levels

The system is composed of 5 levels. The tags default to level 0. When a user moves close to a marker, the level increases by one unit. The perceptible zone in the range of proximity gives an approximate RSSI value of -50. If the user moves away from the proximity area, the RSSI value is less than -50, resulting in a reduction in the level. When a greater level if reached, it is possible to conclude that the user has remained close to the marker, and the computer will be turned on.

On the other hand, reaching an initial level of 0 means that the user has moved a significant distance away from the workspace, and the computer is turned off.

4 Results and Conclusions

This document has presented a proximity detection prototype based on the use of ZigBee technology. The prototype is notably superior to existing technologies using Bluetooth, infrareds or radiofrequencies, and is highly efficient with regards to detection and distance. Additionally, different types of situations in a work environment were taken into account, including nearby computers, shared computers, etc.

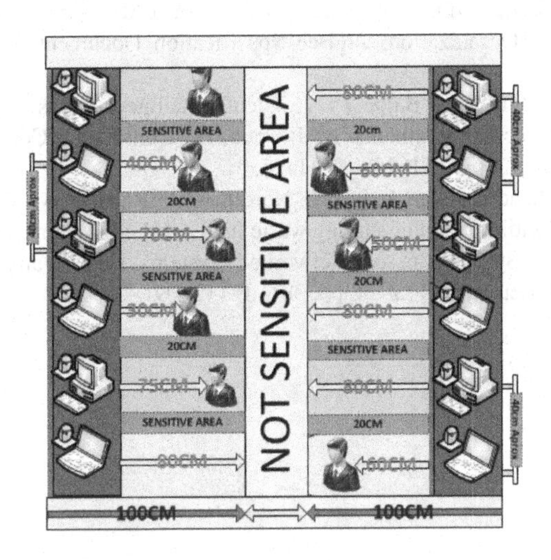

Fig. 6 Diagram of a work environment

Figure 6 represents a possible distribution of computers and laptops in a real office environment, separated by a distance of 2 meters. The activation zone is approximately 90cm, a distance considered close enough to be able to initiate the activation process. It should be noted that there is a "Sensitive Area" in which it is unknown exactly which computer should be switched on; this is because two computers in close proximity may impede the system's efficiency from switching on the desired computer. Tests demonstrate that the optimal distance separating two computers should be at least 40cm.

The prototype is specifically oriented to facilitate the integration of people with disabilities into the workplace. The detection and identification of a user makes it possible to detect any special needs, and for the computer to be automatically adapted for its use. The video in [3] confirms the operation of the prototype.

Acknowledgements. This project has been supported by the Spanish CDTI. Proyecto de Cooperación Interempresas. IDI-20110343, IDI-20110344, IDI-20110345. Project supported by FEDER funds.

References

1. Huang, Y., Pang, A.: A Comprehensive Study of Low-power Operation in IEEE 802.15.4. In: Preceeding of the 10th ACM Symposium on Modeling, Analysis and Simulation of Wireless and Mobile Systems, Chaina, Crete Island, Greece (2007)
2. Singh, C.K., et al.: Performance evaluation of an IEEE 802.15.4 Sensor Network with a Star Topology (2008)
3. Universidad Pontificia de Salamanca. (En línea) (2011), `http://www.youtube.com/watch?v=9iYX-xney6E`
4. Lieberman, P.: Wake on LAN Technology, White paper (2011), `http://www.liebsoft.com/pdfs/Wake_On_LAN.pdf`
5. ZigBee Standards Organization: ZigBee Specification Document 053474r13. ZigBee Alliance (2006)
6. Tapia, D.I., De Paz, Y., Bajo, J.: Ambient Intelligence Based Architecture for Automated Dynamic Environments. In: Borrajo, D., Castillo, L., Corchado, J.M. (eds.) CAEPIA 2007, vol. 2, pp. 151–180 (2011)
7. Nedevschi, S., Chandrashekar, J., Liu, J., Nordman, B., Ratnasamy, S., Taft, N.: Skilled in the art of being idle: reducing energy waste in networked systems. In: Proceedings of the 6th USENIX Symposium on Networked Systems Design and Implementation, Boston, Massachusetts, April 22-24, pp. 381–394 (2009)

Building Context-Aware Services from Non-context-aware Services

Ichiro Satoh

Abstract. This paper presents a framework for providing context-aware services. It supports the separation of services and context, so that application-specific services can be defined independently of any contextual information. It also provides two mechanisms. The first is to enable non-context-aware services to be used as context-aware services. The second is to enable context-aware services to be defined independently on any contextual information. The framework is useful in the development of software for non-context-aware services in ubiquitous computing environments. Our early experiments proved that it enabled us to reuse JavaBeans components as context-aware services without having to modify the components themselves.

1 Introduction

Context-aware services are still one of the most typical applications of ambient computing. They are provided to users with services according to their contexts, e.g., users, locations, and time. Software for context-aware services is assumed to support contextual information or processing inside it. However, such software cannot be reused for other contexts. Furthermore, although there have been numerous non-contextual services, including Web services and components for Java 2 EE, they cannot be directly used in context-aware services.

This paper addresses the reusability of software for context-aware services or non-context-aware services. Our framework consists of two mechanisms. The first is based on the notion of *containers*, used in enterprise software components to emulate the execution environments that the components want to receive. The second is called *connectors*, which loosely bind software for defining services and contextual information in the real world.

Ichiro Satoh
National Institute of Informatics
2-1-2 Hitotsubashi, Chiyoda-ku, Tokyo 101-8430, Japan
e-mail: `ichiro@nii.ac.jp`

P. Novais et al. (Eds.): Ambient Intelligence - Software and Applications, AISC 153, pp. 59–66.
springerlink.com

Our framework is constructed as a middleware system for providing services implemented as JavaBeans like Enterprise JavaBeans (EJB), [4][1] It was inspired by our practical requirements in the sense that we have been required to provide context-aware services in public and private environments, such as that at public museums, retail stores, and corporate buildings.

2 Example Scenario

This approach was inspired by our experiment in the development of practical context-aware services in real spaces, e.g., museums and schools. It supports two kinds of context-aware services for visitors in museums.

- *Context-aware annotation services for exhibits:* Most visitors to museums want annotations on the exhibits in front of them, because they lack sufficient breath of knowledge about them. Their knowledge and experiences are varied so that they tend to be become puzzled (or bored) if the annotations provided to them are beyond (or beneath) their knowledge or interest. Our experiments provided visitors with annotation services that are aware of the users and their current locations.
- *Post-It services on exhibits:* Social network services (SNS), e.g., Facebook and Google+, enables users to make comments on other users' posts or express their positive impressions on the posts. Like SNS, some visitors want to leave their impressions or recommendations on exhibits. Other visitors want to read the impressions and recommendations. Our experiments provided services that enabled visitors to comment on the exhibits and other visitors to read the comments while they were close to the exhibits.

The contexts of the both services are different, but their application logic was similar in that it provided contents to users. Our goal was to enables context-aware services whose application logic was common to be implemented by the same software by explicitly specifying the context in which the services should be activated.

3 Approach

We introduce two mechanisms, called *containers* and *connectors*, into context-aware services.

3.1 Context-Aware Container

Modern enterprise architectures, e.g., Enterprise JavaBeans (EJB), [4], and .Net architecture [5] have employed the notion of containers to separate business components from system components. The original notion has enabled key functionality

[1] Since the framework is aimed at context-aware services, it does not support several system issues, e.g., transactions, used in middleware for enterprise systems.

such as transactions, persistence, or security to be transparently added to application at the time of deployment rather than having to implement it as part of the application. The notion leads to increased reusability and interoperability of business components. We used the notion to reuse non-context-aware business-software components in context-aware ones. Non-context-aware components are not designed to be used in ubiquitous computing environments, where services appear and disappear arbitrarily and nodes cannot possibly know in advance with which other nodes they will interact. Our container mechanism hide such dynamic environments from the components.

3.2 Context-Aware Connector

This was introduced as a spatial relationship between services and the targets that the services should be provided for, e.g., users, physical entities, and spaces. It deploys services at appropriate computers according to the locations of their targets. For example, when a user moves from location to location, it automatically deploys his/her services at computers close to his/her destination. It enables software components for defining services to specify their placement outside them. The current implementation provides two types of context-aware connectors, as shown in Figure 1.

- If a service declares a *follow* connector for at most one moving target, e.g., physical entity or person, the former is deployed at a computer close to the latter's destination, even when the latter moves to another location.
- If a service declares a *shift* connector for at most one moving target, e.g., physical entity or person, the former is deployed at a computer close to the latter's source.

4 Design and Implementation

Our user/location-aware system to guide visitors is managed in a non-centralized manner. It consists of four subsystems: 1) location-aware directory servers, 2) runtime systems, 3) virtual counterparts, and 4) context-aware containers. The first is responsible for reflecting changes in the real world and the locations of users when services are deployed at appropriate computers. The second runs on stationary computers located at specified spots close to exhibits in a museum. It can execute application-specific components via context-aware containers, where we have assumed that the computers are located at specified spots in public spaces and are equipped with user-interface devices, e.g., display screens and loudspeakers. It is also responsible for managing context-aware connectors. The third is managed by the first and deployed at a runtime system running on a computer close to its target, e.g., person, physical entity, or space. The fourth is implemented as a mobile agent. Each mobile agent is a self-contained autonomous programming entity. Application-specific services are encapsulated within the fourth.

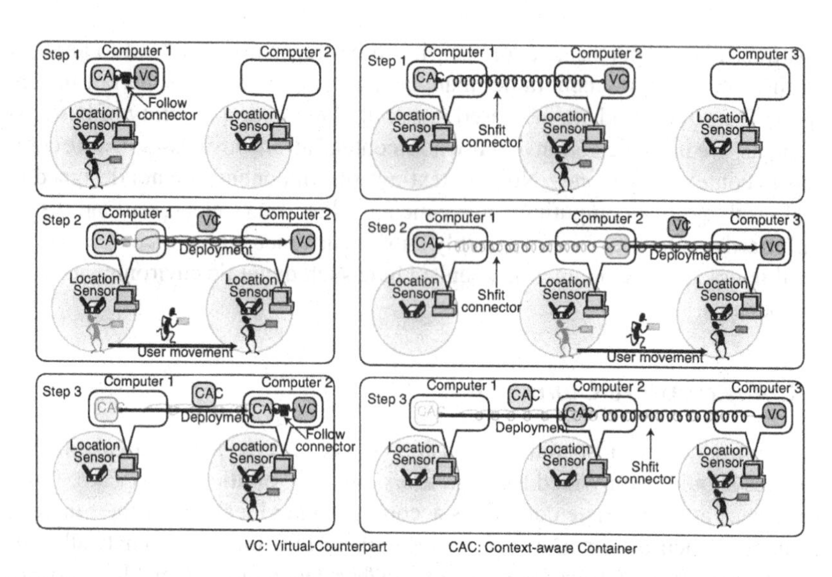

Fig. 1 Context-aware-coupling between virtual-counterparts and services

The system has three unique functions:

- *Virtual-counterpart* is a digital representation of a user, physical entity, or computing device. When its target moves to another location, it is automatically deployed at a computer close to the current location of the target by using location-sensing systems.
- *Context-aware-container* is a customizable wrapper for (non-context-aware) software components, e.g., JavaBeans, for defining application-specific services to use them as context-aware services.
- *Context-aware-connector* is the relationship between the locations of one virtual-counterpart and context-aware container. It deploys or activates the latter at a computer according to its deployment policy (Figure 1).

We assume that virtual-counterparts are managed in the underlying location models. In fact, digital representations in most symbolic location models are directly used as virtual counterparts, where such models maintain the locations of physical entities, people, and spaces as the structure of their virtual counterparts according to their containment relationships in the real world. For example, if a user is in a room on a floor, the counterpart corresponding to the user is contained in the counterpart corresponding to the room and the latter is contained in the counterpart corresponding to the floor. The current implementation support our location model, although the framework itself is not independent of the model[2]. Since the model monitors its underlying location sensing systems, when it detects the movements of physical entities or people in the real world, it reflects the movement on the containment relationships of the virtual counterpart corresponding to the moving entities or people.

[2] Satoh2005

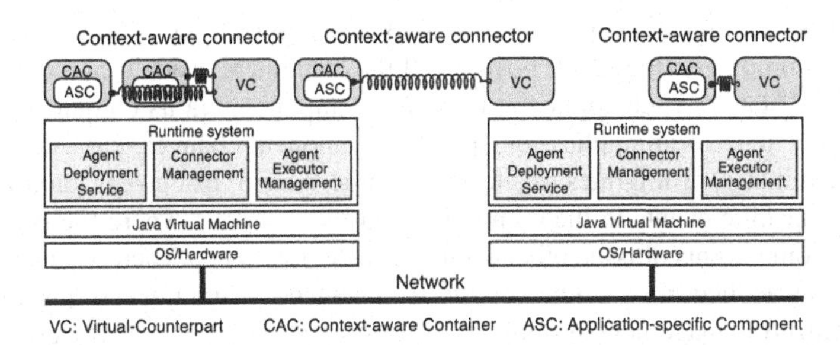

Fig. 2 Service runtime systems

4.1 Context-Aware Container

Each context-aware container is an autonomous programmable entity implemented as a mobile agent. We developed a mobile agent-based emulator to emulate the physical mobility of its target terminal by using the logical mobility of the emulator [9, 10]. It could provide application-level software with the runtime environment compatible to its target device and carry it between computers through networks. Context-aware containers can provide application-specific components with their favorite runtime environments and carry them between computers. They are defined according to types of application-specific components. Each context-aware container in the current implementation is a collection of Java objects and support Java-based components, e.g., JavaBeans and Java Applets.[3]

4.2 Context-Aware Connector

The framework enables each container to have a connector for at most one virtual counterpart. Each connector is activated when its target virtual counterpart moves in the location model according to the movement of the counterpart's target in the physical world. Next, it deploys the container, including a service, at a computer according to its deployment policy, if the computer does not have the service. It activates the service wrapped in the container at the computer. The current implementation supports the two built-in policies explained in Figure 1. Nevertheless, connectors can be extended by overwriting classes for the built-in connectors.

4.3 Runtime System

Each service runtime system is responsible for executing and migrating application-specific components wrapped in context-aware containers to other service runtime systems running on different computers through a TCP channel using mobile-agent technology. It is built on the Java virtual machine (Java VM version 1.5 or later),

[3] The framework itself is not independent of Java and has the potential to support existing services written in other languages.

which conceals differences between the platform architectures of the source and destination computers (Fig. 2). It governs all the containers inside it and maintains the life-cycle state of each application-specific component via its container. When the life-cycle state of an application-specific component changes, e.g., when it is created, terminates, or migrates to another runtime system, its current runtime system issues specific events to the component via its container, where the container may mask some events or issue other events. The deployment of each context-aware container is specified in its connector and is managed by runtime systems without any centralized management system. Each runtime system periodically advertises its address to the others through UDP multicasting, and these runtime systems then return their addresses and capabilities to the runtime system through a TCP channel.

4.4 Current Status

A prototype implementation of this framework was constructed with Sun's Java Developer Kit, version 1.5 or later version. Although the current implementation was not constructed for performance, we evaluated the migration of a context-aware container based on connectors. When a container declares a *follow* or *shift* connector for a virtual-counterpart, the cost of migrating the former to the destination or the source of the latter after the latter has begun to migrate is 88 ms or 85 ms, where three computers over a TCP connection is 32 ms.[4] This experiment was done with three computers (Intel Core 2 Duo 2 GHz with MacOS X 10.6 and Java Development Kit ver.6) connected through a Fast Ethernet network. Migrating containers included the cost of opening a TCP-transmission, marshalling the agents, migrating them from their source computers to their destination computers, unmarshalling them, and verifying security.

5 Application

We constructed and conducted an experiment at the Museum of Nature and Human Activities in Hyogo, Japan, using the proposed framework. The experiment consisted of four spots in front of exhibits, where each spot had an active RFID tag reader. We provided each visitor with an RFID tag. Location-aware directory servers monitored one or more RFID tag readers. When a location-aware directory server detected the presence of an RFID tag in a spot, it instructed the underlying location model to migrated the virtual counterpart corresponding to the visitor attached to the tag to the counterpart corresponding to the spot.

We provided visitors with the two kinds of services discussed in Section 2, i.e., context-aware annotation services for exhibits and post-it services for exhibits. These services were implemented as Javabeans software and their containers were also the same. The software was a multimedia player to display rich text on the current computer. The container for the former service had a follow policy for the

[4] The size of each virtual-counterpart was about 8 KB in size.

counterpart corresponding to a user and the container for the second had a shift policy for the counterpart corresponding to a user.

We offered a GUI system that enabled curators in the museum to customize context-aware connectors. They were required to change the contexts that services should be activated at the experiment because they relocated some exhibits in the room. They could assign either a follow or shift connector to such services wrapped in containers for Javabeans components and they services simultaneously provide two kinds of services. In fact, they could easily customize the contexts through context-aware connectors, because they did not need to know the services themselves.

We did the experiment over two weeks. Each day, more than 80 individuals or groups took part in the experiment. Most visitors answered questionnaires about their impressions on the system. Almost all the participants (more than 95 percent) provided positive feedback on it. As application-specific services could be defined as JavaBeans, we were able to easily test and change the services provided by modifying the corresponding agents while the entire system was running.

6 Related Work

Many researchers have studied software engineering for context-aware services. The Context toolkit was a pioneering work on software engineering issues in context-aware services [1, 8]. It aimed at allowing programmers to leverage off existing building blocks to build interactive systems more easily. It was constructed as libraries of widgets for GUI. However, since it was only designed for context-aware services, it did not support the reuse of software for non-context-aware services.

Ubiquitous computing defines a new domain in which large collections of heterogeneous devices are available to support the execution of applications. These applications become dynamic entities with multiple input and output alternatives. As a result, it is difficult to predict in advance the most appropriate configuration for application as has been discussed by several researchers [7, 11]. There have been many attempts to construct software component technology for ubiquitous computing [2, 6]. Several researchers have done studies modeling context-awareness in the literature of software engineering [3]. However, there have been no silver bullet as other systems thus far.

7 Conclusion

We constructed a framework for providing context-aware services. It supported the separation of services and context, so that application-specific services could be defined independently of any contextual information. It also provided two mechanisms, called *context-aware containers* and *context-aware connectors*. The first enabled non-context-aware services to be used as context-aware services to be used as context-aware services. The second enabled context-aware services to be defined independently of any contextual information.

References

1. Abowd, G.D.: Software Engineering Issues for Ubiquitous Compuitng. In: Proceedings of International Conference on Software Engineering (ICSE 1999), pp. 75–84. ACM Press (1999)
2. Areski, F., Christophe, G., Philippe, M.: A Component-based Software Infrastructure for Ubiquitous Computing. In: Proceedings of the 4th International Symposium on Parallel and Distributed Computing, vol. 8, pp. 183–190. IEEE Computer Society (2005)
3. Henricksen, K., Indulska, J.: Developing Context-Aware Pervasive Computing Applications: Models and Approach. Pervasive and Mobile Computing 2 (2005)
4. Kassem, N.: Designing Enterprise Applications with the Java 2 Plaform, Sun J2EE Blueprints, Sun Microsystems (2000), http://java.sun.com/j2ee/download.html
5. Micorsoft Corp.: The .NET Architecture, Microsoft Corporation (2000), http://www.microsoft.com/net/
6. Martin, M., Umakishore, R.: UbiqStack: a taxonomy for a ubiquitous computing software stack. Personal Ubiquitous Computing 10(1), 21–27 (2005)
7. Roman, M., Al-muhtadi, J., Ziebart, B., Campbell, R., Mickunas, M.D.: System Support for Rapid Ubiquitous Computing Application Development and Evaluation. In: Proceedings of Workshop on System Support for Ubiquitous Computing (UbiSys 2003) (2003)
8. Salber, D., Dey, A.K., Abowd, G.D.: The Context Toolkit: Aiding the Development of Context-Enabled Applications. In: Proceedings of International Conference on Computer-Human Interaction (CHI 1999), pp. 15–20. ACM Press (1999)
9. Satoh, I.: A Testing Framework forMobile Computing Software. IEEE Trasaction of Software Engineering 29(12), 1112–1121 (2003)
10. Satoh, I.: Software Testing for Wireless Mobile Computing. IEEE Wireless Communications 11(5), 58–64 (2004)
11. Scholtz, J., Consolvo, S., Scholtz, J., Consolvo, S.: Towards a discipline for evaluating ubiquitous computing applications, National Institute of Standards and Technology (2004), http://www.itl.nist.gov/iad/vvrg/newweb/ubiq/docs/1scholtzmodified.pdf

S-CRETA: Smart Classroom Real-Time Assistance

Koutraki Maria, Efthymiou Vasilis, and Antoniou Grigoris

Abstract. In this paper we present our work in a real-time, context-aware system, applied in a smart classroom domain, which aims to assist its users after recognizing any occurring activity. We exploit the advantages of ontologies in order to model the context and introduce as well a method for extracting information from an ontology and using it in a machine learning dataset. This method enables real-time reasoning on high-level-activities recognition. We describe the overview of our system as well as a typical usage scenario to indicate how our system would react in this specific situation. An experimental evaluation of our system in a real, publicly available lecture is also presented.

Keywords: AmI, Smart Classroom, Activity Recognition, Context modeling.

1 Introduction

In a typical classroom, a lot of time and effort is sometimes spent on technical issues, such as lighting, projector set-up, photocopy distribution etc. This time could be replaced with "teaching time", if all these issues were solved automatically.

There are many Smart Classroom systems that try to change the behavior of an environment in order to improve the conditions of a class. One of them is [1] that focuses on making real-time context decisions in a smart classroom based on information collecting from environment sensors, polices and rules. Another context aware system is [2] that supports ubiquitous computing in a school classroom.

In this paper, we present a system that assists instructors and students in a smart classroom, in order to avoid spending time in such minor issues and stay focused on the teaching process, by also having more studying material at their disposal. To accomplish this, we have taken advantage of the benefits that ontologies and machine learning offer, unlike other similar systems.

Koutraki Maria · Efthymiou Vasilis · Antoniou Grigoris
Foundation of Reasearch and Technology Heraklion
e-mail: {kutraki,vefthym,antoniou}@ics.forth.gr

P. Novais et al. (Eds.): Ambient Intelligence - Software and Applications, AISC 153, pp. 67–74.
springerlink.com

Ontologies play a pivotal role not only for the semantic web, but also in pervasive computing and next generation mobile communication systems. They provide formalizations to project real-life entities onto machine-understandable data constructs [3]. In our system, machine learning algorithms use these data constructs to conceive a higher level representation of the context.

Following the modern needs, we tried to make a system that will be robust and fast enough, to run reliably in real-time conditions. However, we did not want to sacrifice accuracy in favor of speed or vice versa. So we believe that our approach manages to achieve accuracy and execution time that is comparable to or even better than the state-of-the-art systems.

2 Motivating Usage Scenarios

Scenarios in our work express the activities that can take place in a classroom e.g. lecture, exam. In order to identify these activities, a series of simpler events should be applied, in each case. After that, we undertake to assist the particular activity.

A typical scenario is "Student Presentation". A student is giving a lecture or paper presentation in the Smart Classroom. For this scenario many students and teachers may appear in classroom. The classroom's calendar is checked for lecture at this time and personal calendars of all these people are checked for participation in this presentation, in order to know that everyone is in the right Classroom. The lights and projector are turned on. A student stands near the display of presentation. Teachers and students should be seated. After all these minor events "Student Presentation" activity is identified. After the identification of the activity, we assist the presentation by turning off the lights and adapting the presentation file in students' and teachers' devices.

3 Architecture for Building AmI Systems

A simple description of a complete cycle of our system is the following, as depicted in Figure 1:

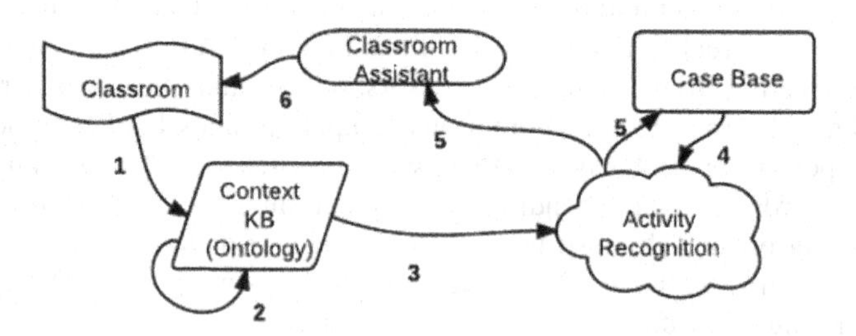

Fig. 1 System Architecture

1. Data from sensors and from services of AMI Sandbox [4] are stored in an ontology. These services provide functionality like localization searching or speech recognition. In our scenario we assume that the localization searching service provide data about students' and teacher's location which are stored into our ontology as well data from lights sensor or data from RFID sensors for people identification.
2. SWRL rules are used for a first level of reasoning, to create *Simple Events*. In "Student Presentation", some simple events are: *student stands near the display of presentation, teacher sits*.
3. The *Simple Events* that occurred within a timeframe are passed to the Activity Recognition System.
4. The Activity Recognition System loads the cases and finds the current activity.
5. The result is written in the case base as a new case and also passed to the *Classroom Assistant* system.
6. Depending on the current activity, the *Classroom Assistant* changes the context. In our scenario turns of the lights and adapt the presentation's file in students' and teachers' devices.

3.1 Modeling Context

In this session we propose a context ontology model for an intelligent university Classroom that responds to students' and teachers' needs. Our context ontology is divided into two hierarchical parts, upper-level ontology and low-level ontologies.

The *Upper Level Ontology* or *Core Ontology* captures general features of all pervasive computing domains. It is designed in a way that can be reused in the modeling of different smart space environments, like smart homes and smart meeting spaces. The Core ontology's context model is structured by a set of abstract entities like Computational Entity, Person, Activity, Location, Simple Event and Environment (Figure 2). All these entities are widely used, except Simple Event entity. Simple Event entity aims to capture knowledge obtained from reasoning on sensors data e.g. 'Projector's status is "on"' or 'Teacher is in front of the smart board'.

The *Low Level* or *Domain-Specific Ontologies* are based on upper level ontology and specified by the domain. In our case the domain is an intelligent classroom in a university campus. Some of the domain-specific ontologies are Person and Location ontology. All of the ontologies are expressed in OWL.

3.2 Reasoning: SWRL

In our implementation we try to transform our scenarios for intelligent classroom into rules. This rule based approach is implemented by using SWRL (Semantic Web Rule Language). The first step is to capture data from sensors and services into the ontology (e.g. status of devices). After that, SWRL rules are applied on these data and save the result into the Simple Event class. Examples of rules are shown below.

Rule 1: *Person:Teacher(?t) ^ Device:SmartBoard(?b) ^ Core:hasRelativeLocation(?t,?b) ^ Core:inFrontOf(?t,?b) ^ Core:SimpleEvent(Teacher in front of the Board) → Core: isActivated(Teacher in front of the Board, "true")*

Rule 2: *Location:SmartClassroom(?c)^Environment(SmartClassroomEnv) ^ Core:hasEnvironment(?c,SmartClassroomEnv)^Core:noiseLevel(SmartClassroomEnv,?noise) ^swrlb:greaterThan(?noise,80) ^ Core:SimpleEvent(High Level Noise) →Core:isActivated(High Level Noise, "true")*

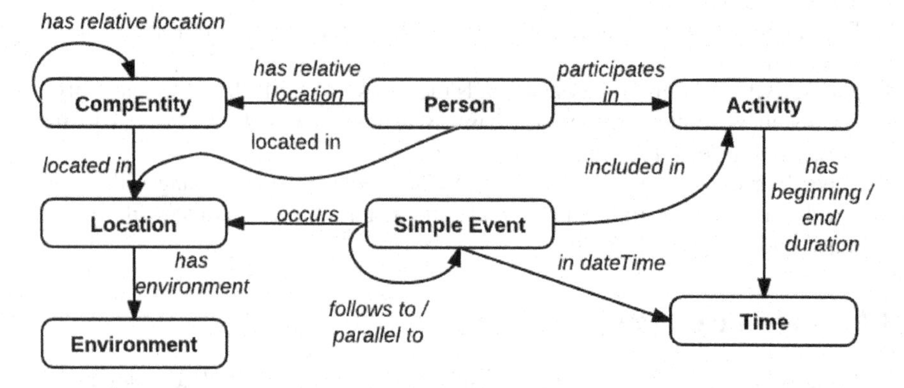

Fig. 2 Core Ontology

3.3 Activity Recognition

For this part of our system, we used (an adjustment of) an activity recognition project, based on Case-Based Reasoning[5], that we are currently developing in parallel, described briefly in chapter 3. After the SWRL rules trigger, the resulting *Simple Events* that occurred within a time frame are sent to the Activity Recognition system, building an unsolved case. One or no activity is then recognized using Bayesian Networks (BNs) and the solved case is added to the case base.

Based on a given set of activities that are to be recognized, a case base is initially created and classified manually. It is essential for this dataset to be a product of real observations and not just random cases. For the implementation of the BNs, we use WEKA[6] with default parameters, so our dataset is stored as *arff*.

Some signal segmentation approaches are described in details in [7]. The one we chose is "not overlapping sliding windows", since this implementation is simpler, faster and activities occurring in the edge of a time window are rare. Especially in a smart classroom, activities usually last longer than other domains and their number is significantly lower. The length of the time window (10 seconds) is chosen based on the nature of the activities and experimental results.

4 Activity Recognition in the Presence of Ontologies

Since ontologies offer ways of expressing concepts and relationships between them, we found it interesting to exploit such expressiveness and assist machine learning.

The most popular Case Based Reasoning (CBR) tool that supports ontologies is JCOLIBRI [8, 9]. As most of the CBR tools do, JCOLIBRI uses the k Nearest Neighbours (kNN) algorithm to classify a new case, so it bares the problems of kNN. Another promising tool, CREEK[10] - recently renamed as AmICREEK – is not publicly available, although some case-evaluation studies have been published [11]. The most recent and similar tool that we found is SituRes[12], a case-based approach to recognizing user activity in a smart-home environment. SituRes is based on one of the publicly available datasets that use PlaceLab [13].

When it comes to real-time activity recognition, there is a need for accuracy as well as speed. Our approach aims to take advantage of the rich expressiveness that ontologies can offer and provide solid answers, using machine learning algorithms, like Support Vector Machines (SVMs) or BNs. The key factor that led to the design of a new system, using machine learning, is the lack of speed observed in the already existing systems that use algorithms such as kNN. Apart from that, the robustness and accuracy of BNs led to a system faster and more accurate than other systems that we are aware of. Coping with sensors usually means missing data and SVMs – our initial choice - lack in this field. BNs are similar in terms of accuracy and outperform SVMs when data are missing, as proven in [14].

The input of the complete system is an ontology with instances, from which the user has to define the terms (classes) that describe the attributes, the term that describes the solution and the term where the cases are stored. For example, consider a simple ontology, where *Case, Activity, Location, Time, Winter, Summer, Autumn, Spring, Indoors, Outdoors, FirstFloor* and *SecondFloor* are classes. As expected, *Winter, Summer, Autumn* and *Spring* are subclasses of *Time*. *Indoors* and *Outdoors* are subclasses of *Location* and *FirstFloor* and *SecondFloor* are subclasses of *Indoors*. In this ontology there should be some *Case* instances like the following:

```
<Case rdf:ID="Case74">
  <has-Activity rdf:resource="#Cooking"/>
  <has-Time rdf:resource="#July"/>
  <has-Location rdf:resource="#Kitchen"/>
</Case>
```

where *July* is an instance of *Summer* and *Kitchen* is an instance of *FirstFloor*. The user should define that terms *Time* and *Location* describe the attributes of the problem, term *Activity* is used to describe the solution and each case is stored as an instance of term *Case*. The suggested solution to the problem of grasping the hierarchy information of an ontology and storing it as attributes, is keeping the whole path of each instance in Boolean values. For example *Case74* of the example above, would be stored as:

Cooking, July,0,0,0,1,Kitchen,1,0,1,0

The Boolean values following *Kitchen* mean that it belongs to *FirstFloor* and *Indoors* and not *SecondFloor* or *Outdoors.* In other words, after an attribute value, we store one Boolean number for each of its subclasses, representing that the subclass belongs or not to the path that leads to the instance value.

In our first version of the activity recognition system, we used ontologies as described above and the results were satisfying enough. For facilitating the evaluation process we developed a simpler but quicker version without using ontologies. In this version, events occurring within a timeframe are sent as plain text (and not within an ontology) to the activity recognition system, and added as a case to the case-base. The accuracy of the simpler version was close to the first version, but time performance was better, as expected.

5 Experimental Evaluation

In the absence of a real dataset for a smart classroom, we decided to create one, in order to evaluate our system. The precision of our system is mostly based on the activity recognition's precision, since everything else is rule based. Therefore we present here the evaluation results for a dataset that we built based on our observations on a publicly available video from a lecture[1]. Our observations – which act as sensor data – include the position of the lecturer, the lighting, the persons that speak etc. In this video the activities observed are 4: lecture with slides, lecture with whiteboard, question and conversation. A 10 fold cross validation based on this dataset is illustrated in Tables 1 and 2. Table 2 can be read as "y activity was actually classified as x activity". It would ideally contain zeros only in the non-diagonal positions. So it appears that no "conversation" was classified correctly. Further evaluation experiments of the activity recognition system, will be presented in a later work.

Total Number of Instances	326	
Correctly Classified Instances	313	96.0123 %
Incorrectly Classified Instances	13	3.9877 %

Table 1 Detailed accuracy by class

TP Rate	FP Rate	Precision	Recall	F-Measure	ROC Area	Class
0.995	0.077	0.959	0.995	0.977	0.994	lecture_slides
0.955	0.008	0.977	0.955	0.966	0.994	lecture_wb
0.955	0.007	0.913	0.955	0.933	0.997	question
0	0	0	0	0	0.816	conversation

[1] http://videolectures.net/mlss08au_hutter_isml/. Part 2

Table 2 Confusion matrix

lecture_slides	lecture_wb	question	conversation	←classified as
208	1	0	0	lecture_slides
4	84	0	0	lecture_wb
1	0	21	0	question
4	1	2	0	conversation

The time performance of our system is based on the performance of SWRL plus the performance of the activity recognition system. As stated in §3.3, we have set a typical reasoning cycle (case) to 10 seconds. This means that for a typical activity recognition dataset we would need around 43000 cases, which correspond to 5 (working) days of data, namely a week. We have reproduced the same 326 cases acquired from the video to create a 43000-case-large dataset, just to simulate the time performance of our system, ignoring accuracy. The average time performance of the activity recognition system that was executed 100 times on this dataset is 0.757401 seconds in a rather outdated machine. Similarly, the average time performance of SWRL rules that was executed 10 times is 0.686215 seconds.

6 Conclusions – Future Work

In this paper, we introduced the use of a real-time AmI system in a smart classroom. We presented how an ontology can be used to model the context in a smart environment and how we can take advantage of this modeling to assist activity recognition.

With some simple scenarios, we illustrated how such a system could assist its users and also provided some experimental results on the performance. Although the first results are promising, we still have some work to do in order to test our system in a real smart classroom environment.

Apart from that, we have already started to work on the Ambient Assisted Living domain and particularly on the assistance of the elderly. We also plan to extend our Activity Recognition system, in order to grasp more information that an ontology can offer, reduce the case-base's size efficiently and finally verify which activities were recognized correctly.

Acknowledgments. We would like to thank Ioannis Hryssakis and Dimitra Zografistou for their support and advice in the preparation of this paper, as well as their eagerness to help whenever needed.

References

[1] O'Driscoll, C., Mohan, M., Mtenzi, F., Wu, B.: Deploying a Context Aware Smart Classroom. In: International Technology and Education Conference. INTED, Valencia (2008)

[2] Leonidis, A., Margetis, G., Antona, M., Stephanidis, C.: ClassMATE: Enabling Ambient Intelligence in the Classroom. World Academy of Science, Engineering and Technology 66, 594–598 (2010)
[3] Krummenacher, R., Strang, T.: Ontology-based Context Modeling. In: Proceedings Third Workshop on Context-Aware Proactive Systems, CAPS 2007 (2007)
[4] Grammenos, D., Zabulis, X., Argyros, A., Stefanidis, C.: FORTH-ICS Internal RTD Programme Ambient Intelligence and Smart Environments. In: Proceedings of the 3rd European Conference on Ambient Intelligence (AMI 2009) (2009)
[5] Aamodt, A., Plaza, E.: Case-Based Reasoning: Foundational Issues, Methodological Variations, and System Approaches. AI Communications 7(1), 39–59 (1994)
[6] Witten, I.H., Frank, E., Hall, M.A.: Data Mining: Practical Machine Learning Tools and Techniques, 3rd edn. Morgan Kaufmann (2011)
[7] Tapia, E.M.: Using Machine Learning for Real-time Activity Recognition and Estimation of Energy Expenditure. Dissertation, Massachusetts Institute of Technology (2008)
[8] Recio-Garcia, J.A.: jCOLIBRI: A multi-level platform for building and generating CBR systems. Dissertation, Universidad Complutense de Madrid (2008)
[9] Recio-Garcia, J.A., Diaz-Agudo, B., Gonzalez-Calero, P., Sanchez-Ruiz-Granados, A.: Ontology based CBR with jCOLIBRI. Applications and Innovations in Intelligent Systems Xiva (2007)
[10] Aamodt, A.: A knowledge-intensive, integrated approach to problem solving and sustained learning. Dissertation, University of Trondheim, Norwegian Institute of Technology, Department of Computer Science, University Microfilms PUB 92-08460 (1991)
[11] Kofod-Petersen, A., Aamodt, A.: Case-Based Reasoning for Situation-Aware Ambient Intelligence: A Hospital Ward Evaluation Study. In: McGinty, L., Wilson, D.C. (eds.) ICCBR 2009. LNCS, vol. 5650, pp. 450–464. Springer, Heidelberg (2009)
[12] Knox, S., Coyle, L., Dobson, S.: Using ontologies in case-based activity recognition. In: Proceedings of FLAIRS 2010, pp. 336–341. AAAI Press (2010)
[13] Intille, S.S., Larson, K., Beaudin, J.S., Nawyn, J., Tapia, E.M., Kaushik, P.: A Living Laboratory for the Design and Evaluation of Ubiquitous Computing Technologies. In: Proceedings of CHI Extended Abstracts, pp. 1941–1944 (2005)
[14] Jayasurya, K., Fung, G., Yu, S., Dehing-Oberije, C., De Ruysscher, D., Hope, A., De Neve, W., Lievens, Y., Lambin, P., Dekkera, A.L.A.J.: Comparison of Bayesian network and support vector machine models for two-year survival prediction in lung cancer patients treated with radiotherapy. Med. Phys. 37, 1401–1407 (2010)

Inferring Ties for Social-Aware Ambient Intelligence: The Facebook Case*

Sandra Servia Rodríguez, Ana Fernández Vilas,
Rebeca P. Díaz Redondo, and José J. Pazos Arias

Abstract. This paper proposes a cloud-based solution to social-aware Ambient Intelligence. The proposal uses the common space of the cloud to gather evidence of social ties among users, turning the social cloud into a bridge between smart spaces that adapt themselves according to, not only the situational information in the physical space, but also to the social sphere of the user. As case instance of the approach, we detail the estimation of social ties from the interaction activity of Facebook users.

1 Introduction

As materialisation of Ambient Intelligence (AmI), smart spaces are sentient and information-rich environments that sense and react to situational information to self-adapt to users' expectations and preferences [3]. Intelligence is embedded around the user (ambient) rather than into user's devices. As intelligent environments mature, they are expected to be helpers in the human society, becoming essential for them to have sophisticated social abilities (communicate, interact) but also sophisticated social aware behaviour related to the user and her social ties. The evolution of AmI into the social sphere poses new challenges to the usual requirements of *personalisation*, *adaptation* and *anticipation* in smart environments. AmI in the social era can take the form of a *socialised*, *empathic* or *conscious* system [2], but also the common personalised, anticipatory and adaptive requisites for ambient intelligence turn into social-aware ones.

Sandra Servia Rodríguez · Ana Fernández Vilas · Rebeca P. Díaz Redondo ·
José J. Pazos Arias
Department of Telematics Engineering, Escuela de Ingeniería de Telecomunicación,
University of Vigo, Spain
e-mail: {sandra,avilas,rebeca,jose}@det.uvigo.es

* Work funded by the Ministerio de Educación y Ciencia (Gobierno de España) research project TIN2010-20797 (partly financed with FEDER funds), and by the Consellería de Educación e Ordenación Universitaria (Xunta de Galicia) incentives file CN 2011/023 (partly financed with FEDER funds).

P. Novais et al. (Eds.): Ambient Intelligence - Software and Applications, AISC 153, pp. 75–83.
springerlink.com

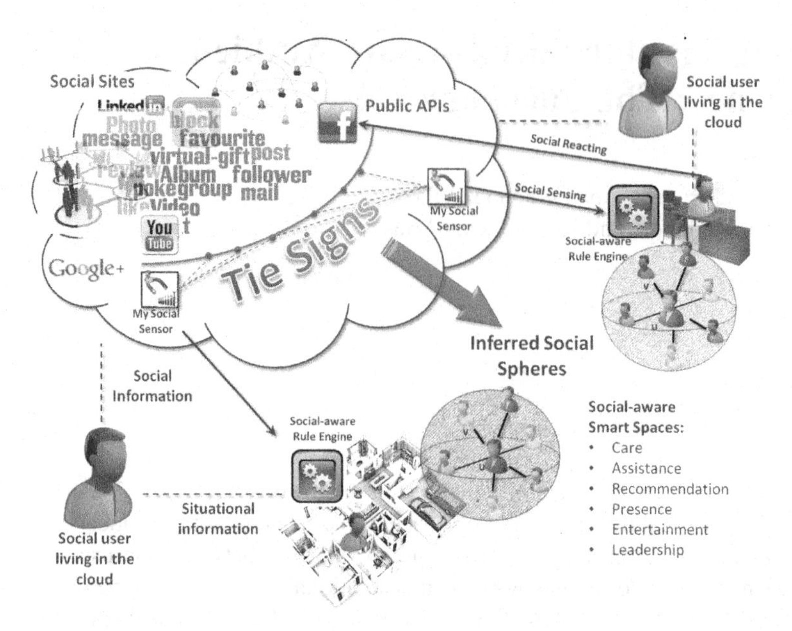

Fig. 1 Social-aware Ambient Intelligence

Ambient intelligence in smart spaces detects signs that trigger a personalised, adaptive and anticipatory action. These signs are accessible to the smart space engine through embedded devices. However, the social dimension goes far beyond many physical sensors at the user's location and being aware of how the user relates with others comes into play. Consequently, it is necessary to establish a global social-aware model, deployed in the cloud, which have updated information on ties between users, and integrate such information with the usual context data.

With this scenario in mind, the project CLOUDIA proposes (1) a social P2P model in the cloud where peers are socially linked and (2) a mechanism for inferring these social ties between users from their interactions in the cloud. Obviously, this inferring engine only use the users' information they have made accessible in different social networks (pillars of interaction in the cloud), obeying the usual privacy and security policies. We have selected Facebook to put into practice our approach of gathering users' interaction activity since it is the largest social network [12] with over 800 million active users [1]. Anyway, the procedure can be easily generalised to any social networking site with a public API.

Our Approach to Social-Aware AmI. Our social ambient intelligence proposal, Figure 1, provides a model that can accommodate the growing number of users living in the cloud, using services like *Google Chat*, *YouTube* or *Twitter*. The concept of tie strength was introduced by Granovetter [8], who defines it as a function of duration, emotional intensity, intimacy and exchange of services from which ties are split into 'strong' and 'weak' ties. In the past thirty years, many attempts have been made to find valid indicators and predictors of tie-strength, such as frequency of interaction [5] [10], duration [5] [10] or intimacy [5] [10] [8]. In the case of social networks online, some proposals [7] are supported by the information kept in the user's profile (age, political ideals, distance between hometowns, for instance). However, we consider that other signs left by users in the cloud, mainly different modes of interaction, are more reliable. In fact, [9] concludes that attribute-based features (in profiles) are the least representative, being the transactional features (wall-postings, picture posting and groups, for instance) the most outstanding features to predict tie strength in Facebook.

Since a social-aware ambient intelligent environment reacts not only to situational information but also to social information, we need to provide a bidirectional communication mechanism between the cloud and the social smart space. We propose a social sensor service, *My Social Sensor* in the Figure 1, which lives in the cloud and is in charge of monitoring and processing evidences of relationships to build up the user's social sphere. This information is used by the smart space to react accordingly: not only with typical local actions (like light switching), but also with actions in the cloud. So, reversely, also the user's behaviour in the smart space can affect the social dimension in the cloud through social-able devices in the smart space which, for instance, update the status information on a social site from the situational information at home. In this context, future scenarios with social-aware devices at home might detect loneliness, for instance, and automatically warn the user's friends, or a photo-frame might select the photo sequence according to the owner's social sphere.

2 Tie Signs: The Facebook Case

On the premise *the more interaction between two users, the more tie strength*, we have developed a Facebook application[1] that extracts user's activity in Facebook and infers the closeness between a target user, u, and one of this friends, v. Since u probably takes advantage from the Facebook facilities to communicate to v (private messages, wall-posts, photos and videos uploads, etc.), we use all this interactions as signs to built a model that calculates the tie strength between u and v, from the u's perspective: $TS_u(v)$. Please, note that this subjective point of view surely cause that the tie strength from the v's perspective, $TS_v(u)$, is different.

After a detailed analysis of Facebook features and how users interact and communicate, we have identified the interaction signs whose mathematical notation is shown in the following tables:

[1] Using the OAth2.0 protocol, our application requires the target user grants a set of privileges that are explicitly required when joining the application.

Signs	$x \to y$
Wall-posts	$P(x,y) = \{p_1(x,y), p_2(x,y), \ldots\}$
Private messages	$PM(x,y) = \{pm_1(x,y), \ldots\}$
Comments	$C(x,y) = \{c_1(x,y), c_2(x,y), \ldots\}$
Likes	$L(x,y) = \{l_1(x,y), l_2(x,y), \ldots\}$

Signs	***x*'s elements where *y* is tagged**
Photos	$PH(x,y) = \{ph_1(x,y), ph_2(x,y), \ldots\}$
Videos	$VD(x,y) = \{vd_1(x,y), vd_2(x,y), \ldots\}$

Signs	***x*'s Public signs**	***x*'s Secret/Private signs**
Belonging Groups	$G_p(x) = \{g_{p_1}(x), g_{p_2}(x), \ldots\}$	$G_s(x) = \{g_{s_1}(x), g_{s_2}(x), \ldots\}$
Events Attendance	$EV_p(x) = \{ev_{p_1}(x), ev_{p_2}(x), \ldots\}$	$EV_s(x) = \{ev_{s_1}(x), ev_{s_2}(x), \ldots\}$

3 Tie Strength Inference

This paper only focuses on inferring tie strength indexes between Facebook friends, so we only analyse how to assess the closeness u perceives about his relationship with v: $TS_u(v) \in [0,1]$. To obtain the index value, we propose the following function:

$$f(x) = \begin{cases} 0 & \text{if } 0 \leq x \leq \frac{\bar{x}^2}{x_{max}} \\ \frac{\ln(\frac{x_{max}}{\bar{x}^2}x)}{\ln(\frac{x_{max}^2}{\bar{x}^2})} & \text{if } \frac{\bar{x}^2}{x_{max}} < x \end{cases} \tag{1}$$

being $\bar{x}$ and x_{max} the mean and maximum value, respectively. So, $f(x)$ is close to 1 if $x > \bar{x}$, close to 0 if $x < \bar{x}$ and, finally, close to 0.5 if $x \sim \bar{x}$; exactly the behaviour we are looking for. With this function as base, Section 3.1 shows our approach to calculate the tie strength index. However, as life itself, tie strength should be a dynamic index reflecting that old interactions are progressively less important and, so, have less influence in the index calculation. Additionally, some signs' influence vanishes as the number of participants increase, so we have also add the concept of *relevance*. Section 3.2 shows as time and relevance are taken into account in the index calculation.

3.1 Tie Strength Calculation

We propose obtaining the strength index of the tie between u and v, from the u's perspective, as a weighted addition of three kind of interactions: (1) *on-line*, $TS_u|_o(v)$; (2) *physical*, $TS_u|_p(v)$; and (3) *interest-based*, $TS_u|_i(v)$:

$$TS_u(v) = \beta \cdot TS_u|_o(v) + \gamma \cdot TS_u|_f(v) + (1 - \beta - \gamma) \cdot TS_u|_i(v) \tag{2}$$

where β and γ depend on how user uses Facebook facilities.

On-line Interactions ($TS_u|_o(v)$). Under this name we include those signs that happen exclusively in the Facebook world and do not require a previous face-to-face contact: wall-posts, comments, likes and private messages. We define two subsets: addressed-signs and open-signs. The former draw together the interactions that v explicitly sends to u –a private message, for instance; whereas the latter are those interactions without an explicit receiver –any like, for example. So, $TS_u|_o(v)$ is obtained as follows:

$$TS_u|_o(v) = \alpha \cdot f(x_d(v,u)) + (1-\alpha) \cdot f(x_o(v,u)) \tag{3}$$

where $x_d(v,u) = |P(v,u)| + |P(u,v)| + |PM(v,u)| + |PM(u,v)|$ is the number of addressed-signs and $x_o(v,u) = |C(v,u)| + |C(u,v)| + |L(v,u)| + |L(u,v)|$ the number of open-sings. Since α reflects the importance of addressed-signs, that we consider is significantly more relevant than open-signs, it should be more than 0.5.

Face-to-face Interactions ($TS_u|_f(v)$). This contribution reflects any interactions showing a previous physical contact between u and v, so it is obtained as follows:

$$TS_u|_f(v) = f(x) \tag{4}$$

where $x(u,v) = |PH(u,v)| + |VD(u,v)|$ denotes the number of u's photos where v is tagged.

Interest-Based Interactions ($TS_u|_i(v)$). This contribution assesses the common interests that u and v have explicitly shown. In the Facebook universe this may be done by subscribing to a group as well as accepting an event invitation. Thus, it is obtained as follows:

$$TS_u|_i(v) = \alpha \cdot f(y_d(v,u)) + (1-\alpha) \cdot f(y_o(v,u)) \tag{5}$$

where $y_d(v,u) = |G_s(u) \cap G_s(v)| + |EV_s(u) \cap EV_s(v)|$ is the number of addressed sings (private and secret groups and events) and $y_o(v,u) = |G_p(u) \cap G_p(v)| + |EV_p(u) \cap EV_p(v)|$ the number of open-signs (public groups and events); α, since has the same meaning than in Equation 3, should have the same value and be always over 0.5.

3.2 Impact of Time and Relevance

None all Facebook signs, even belonging to the same kind, should have the same relevance in the index calculation. For instance, being tagged together in a five-people photo it is clearly more relevant than being tagged together in a twenty-people photo; at least, it may be assumed that in the first case the situation entails more closeness. So, some signs' relevance vanishes as the number of participants increase. For time we adopt the same pattern: relevance vanishes as time goes by. Thus, we propose to modify the previous equations by using the following decreasing function:

$$d(x) = e^{-\mu \cdot x} \tag{6}$$

where μ represent the strength of the slope, i.e. the velocity to *vanish* signs' importance: μ_r for relevance and μ_t for time.

Relevance Impact. This aspect only affects physical and interest-based contributions in Equation 2 (photos, videos, events and groups). Physical contribution is obtained by:

$$TS_u|_f(v) = f(x(v,u)), \text{ where } x(v,u) = \sum_{\forall j \in PH(u,v)} d(|tags_j|) + \sum_{\forall j \in VD(u,v)} d(|tags_j|)$$

being $|tags_j|$ de number of tags in the j-picture (or video) and $d(|tags_j|)$ the result of applying Equation 6. To obtain interest-based index, Equation 5, we use the following contributions:

$$y_d(v,u) = \sum_{\forall j \in (G_s(u) \cap G_s(v))} d(|users_j|) + \sum_{\forall j \in (EV_s(u) \cap EV_s(v))} d(|users_j|)$$

$$y_o(v,u) = \sum_{\forall j \in (G_p(u) \cap G_p(v))} d(|users_j|) + \sum_{\forall j \in (EV_p(u) \cap EV_p(v))} d(|users_j|)$$

being $|users_j|$ the number of users that are expected to attend j-event or are subscribed in j-group, and $d(|users_j|)$ the result of applying Equation 6.

Time Impact. Time, however, affects all Facebook signs: the older an interaction is, the lower its weight should be. Thus, applying the decreasing function, the contributions to Equations 3, 4 and 5 to calculate $TS_u(v)$ are as follows, being $d(t_j)$ the result of applying Equation 6 to the time of the latest updated of j-Facebook sign:

$$x_d(v,u) = \sum_{\forall j \in P(u,v)} d(t_j) + \sum_{\forall j \in P(v,u)} d(t_j) + \sum_{\forall j \in PM(u,v)} d(t_j) + \sum_{\forall j \in PM(v,u)} d(t_j)$$

$$x_o(v,u) = \sum_{\forall j \in C(u,v)} d(t_j) + \sum_{\forall j \in C(v,u)} d(t_j) + \sum_{\forall j \in L(u,v)} d(t_j) + \sum_{\forall j \in L(v,u)} d(t_j)$$

$$x(v,u) = \sum_{\forall j \in PH(u,v)} d(|tags_j|) \cdot d(t_j) + \sum_{\forall j \in VD(u,v)} d(|tags_j|) \cdot d(t_j)$$

$$y_d(v,u) = \sum_{\forall j \in (G_s(u) \cap G_s(v))} d(|users_j|) \cdot d(t_j) + \sum_{\forall j \in (EV_s(u) \cap EV_s(v))} d(|users_j|) \cdot d(t_j)$$

$$y_o(v,u) = \sum_{\forall j \in (G_p(u) \cap G_p(v))} d(|users_j|) \cdot d(t_j) + \sum_{\forall j \in (EV_p(u) \cap EV_p(v))} d(|users_j|) \cdot d(t_j)$$

4 Experimental Evaluation

We have focused our tests on three stereotyped Facebook users: (1) users having many friends that usually interacts with only a few (our instance is user A having 130 friends, average Facebook user [1]); (2) users having only a few close friends and interacting with all of them (our instance is user B having 11 friends); and (3) users having a few friends with which hardly interact (our instance is user C having 62 friends). We use data from a limited set of Facebook users who joined Facebook before December of 2008 and who often access it.

After several analysis and noticing that considered users have similar use patterns, we have decided that the importance of the directed addressed-signs (α) is

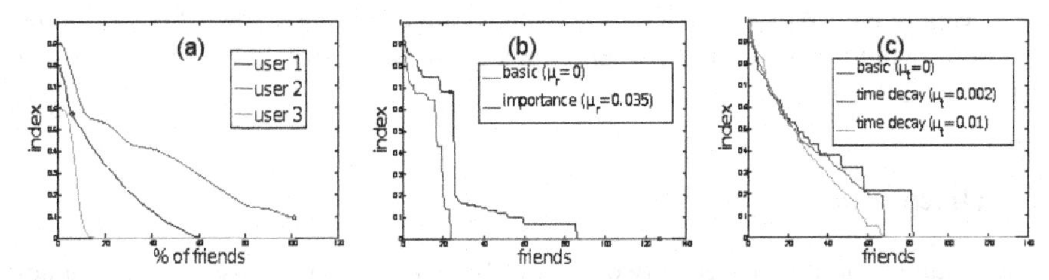

Fig. 2 (a) $TS_A(x)$, $TS_B(x)$ and $TS_C(x)$; (b) Relevance influence in $TS_A(x)$ (only interest-based interactions); (c) Time influence in $TS_A(x)$ (only online interactions)

5 times greater than the opened ones, as well as 60% is the weight for online interactions, 25% for physical interactions and 15% for interest-based interactions. Under this conditions, the tie strength index for friends of users A, B and C is shown in Figure 2. As expected, the index is greater than 0.5 in, at most, the 23% of the considerate cases, coinciding with [12] statement (the large majority of interactions occur only across a small subset of their social links). Besides, the more spread the allocation attention across her friends (or not friends) is, the less difference exists among their indexes. For example, user B, who spreads her interactions among her friends more uniformly than the others, has a lower slope. For example, X is a very active user and has the same interaction signs with her friends A and B, however $TS_A(C) = 0.58$ and $TS_B(C) = 0.009$, as expected. Thus, the tie strength index depends on how the allocation attention across her friends is: high for A (A does not pay attention to many of her friends), whereas for C is much lower, since C spreads her attention more uniformly among her friends.

We have considered that signs' relevance should lose half of their value when the amount of people in the photo, video, group or event is more than 20 users; whereas signs' importance should lose half of their value if they happened two months before the index is calculated.

Signs' Relevance: We have focused on user A, the closest to the average user [1], and the variation of the index over the user's friends is shown in the Figure 2. Under the vanishing conditions previously mentioned (more than 20 people in the sing entails it loses half its value) the forgotten parameter is $\mu_r = 0.035$. According to our assessment, when relevance is not considered, the index is null for the 35% of A's friends, whereas it is taken into account, this percentage rises to 80%. This is interesting because the majority of the groups or events in Facebook have about 5000 attendees, which obviously vanishes the tie between our user and her friend. For example, the user B has an basic index of 0.68, while her value drops to 0 when the relevance is considered.

Time Influence: Focusing on the same user A, Figure 2 shows how time affects the index calculation. If signs loses half of their value after two months of having happened, it entails that $\mu_t = 0.01$, whereas a year of tolerance entails that $\mu_t = 0.002$. Our results shown that the number of A's friends with a null index is greater when time is considered. Besides, the lower μ_t is, the more friends have a null index.

It is keeping with Wilson's article [12], which indicates that the lower the size of the temporal window in which the interactions happened is, the lower the number of friends that the user interacts with is.

5 Discussion

Initial studies on interaction networks have brought great insights into how an activity network is structurally different from the social network. Remarkably, Wilson et al. [12] and Viswanath et al. [11] study the evolution of activity between Facebook users from data obtained using crawlers which only consider Wall post and photo comments. Backstrom et al. [4] study how Facebook users allocate attention across friends but they take in account also messages and comments. Unfortunately, this work cannot be applied to the deployment of an online system since they obtain the information manually from Facebook profiles and not from automated API queries. Our approach gets the most out Facebook API and takes into account all information available with the user's privileges, which is relevant from the point of view of interaction. Regarding the application of the inferred social ties, there are several works that focus their attention on enhanced-social applications ([12] for spam filters, [6] for trust-based recommendation). This paper is, to our knowledge, the first proposal which attacks the problem of social-aware Ambient Intelligence.

Although, in this paper, the tie strength is only based on a Facebook interactions between friends, the proposed formulation may also be used to obtain the tie strength between any two Facebook users, not necessarily friends[2]. This index may be interesting because, even at first sight it is expected for two Facebook friends to have a stronger tie than two non-friends, statistics show Facebook users only regularly relate with a small subset of her 130 friends, on average [1]. Thus, it is perfectly possible for her to have more interaction with a non-friend than with one of her friends, which must not be ignored to obtain users' the social sphere. Also, we consider that the more common friends (friends with strong ties) two users share, the stronger their tie is. For that reason, the more number of common friends in an interaction, the higher its relevance is. So, we plan to take into account the common real friends to calculate tie strength between two users and, likewise, to get a more accurate model of the impact of relevance. Besides, we are currently working on gathering interaction signs form other publicly accessible social sites and, even more, using also evidence on the situational information of the user in the smart spaces.

[2] Please, note that some of the signs, like wall-post are only available for friends, so the absence of these contributions entails a reduction in the tie strength.

References

1. Facebook statistics,
   https://www.facebook.com/press/info.php?statistics
2. Aarts, E., de Ruyter, B.: A new research perspectives on ambient intelligence. Journal of Ambient Intelligence and Smart Environments 1(1), 5–14 (2009)
3. Al-Muhtadi, J., Ranganathan, A., Campbell, R., Mickunas, M.: Cerberus: a context-aware security scheme for smart spaces. In: Proceedings of the First IEEE International Conference on Pervasive Computing and Communications, pp. 489–496 (2003)
4. Backstrom, L., Bakshy, E., Kleinberg, J., Lenton, T., Rosenn, I.: Center of attention: How facebook users allocate attention across friends. In: Proceedings of the 5th International AAAI Conference on Weblogs and Social Media (2011)
5. Blumstein, P., Kollock, P.: Personal relationships. Annual Review of Sociology (1988)
6. Chen, W., Fong, S.: Social network collaborative filtering framework and online trust factors: a case study on facebook. In: The 5th International Conference on Digital Information Management (ICDIM 2010), Thunder Bay, Canada, pp. 266–273 (2010)
7. Gilbert, E., Karahalios, K.: Predicting tie strength with social media. In: Proceedings of the 27th International Conference on Human Factors in Computing Systems on Human Factors in Computing Systems, pp. 211–220. ACM (2009)
8. Granovetter, M.S.: The strength of weak ties. American Journal of Sociology (1973)
9. Kahanda, I., Neville, J.: Using transactional information to predict link strength in online social networks. In: International AAAI Conference on Weblogs and Social Media (2009)
10. Perlman, D., Fehr, B.: The development of intimate relationship. Sage Publications, Inc. (1987)
11. Viswanath, B., Mislove, A., Cha, M., Gummadi, K.: On the evolution of user interaction in facebook. In: Proceedings of the 2nd Workshop on Online Social Networks, pp. 37–42 (2009)
12. Wilson, C., Boe, B., Sala, A., Puttaswamy, K.P., Zhao, B.Y.: User interactions in social networks and their implications. In: Proceedings of the 4th ACM European Conference on Computer Systems, Nuremberg, Germany (2009)

References

A Model-Driven Approach to Requirements Engineering in Ubiquitous Systems

Tomás Ruiz-López, Carlos Rodríguez-Domínguez,
Manuel Noguera, and María José Rodríguez

Abstract. Non-Functional Requirements (NFRs) are of paramount importance for the success of Ubiquitous Systems. However, existing methods and techniques to engineer these systems lack support in their specific and systematic treatment. In this paper, a specification technique and several models are introduced to deal with NFRs paying special attention to those particulary related to the features of Ambient Intelligence (AmI) and Ubiquitous Computing (UC). A Model-Driven approach is followed in order to enable the derivation of software designs for such systems. To this end, formal models and methods are defined, as well as an evaluation procedure to be applied, which aims to help designers to select the most appropriate solutions towards the satisfaction of quality attributes.

1 Introduction

Non-Functional Requirements (NFRs) are key to the success of Software Projects [3]. Their fulfillment determines the quality of the system and allows designers to have a criterion to decide between different choices. This is also true in Ubiquitous and Ambient Intelligence (AmI) Systems, where some of these features are required in most situations; scalability, robustness, accuracy, fault-tolerance or usability are just some of them with crucial importance in this area.

However, most of the existing approaches to Requirements Engineering for Ubiquitous Systems mainly focus on functional features and do not provide a systematic treatment of NFRs [5, 14, 15]. Some other pieces of work do provide such a systematic treatment [3], but lack support to the particular features of AmI and Ubiquitous Systems, namely context-awareness, adaptivity or heterogeneity of situations, among others. Thus, new methods and techniques are needed in the Requirements Engineering area in order to be able to deal with NFRs in Ubiquitous Systems.

Tomás Ruiz-López · Carlos Rodríguez-Domínguez · Manuel Noguera · María José Rodríguez
Department of Software Engineering, University of Granada
C/ Periodista Daniel Saucedo Aranda s/n, 18.014 Granada, Spain
e-mail: {tomruiz,carlosrodriguez,mnoguera,mjfortiz}@ugr.es

P. Novais et al. (Eds.): Ambient Intelligence - Software and Applications, AISC 153, pp. 85–92.
springerlink.com

On the other hand, Model-Driven Development [10] is a actual approach to software development in which systems are specified in terms of high level models. Subsequently, these models are refined and transformed into more specific ones and, eventually, code is automatically generated. This approach provides clear benefits in terms of error-prone software design, ease of maintenance and fast development.

In this paper we introduce a Model-Driven approach to engineering Ubiquitous Systems (MD-UBI) in terms of the Software Process Engineering Metamodel (SPEM) [11]. This process comprises the whole lifecycle of software development, ranging from Requirements Engineering to Deployment and Maintenance, and it is based on the Rational Unified Process [8]. Hereby, we focus on the Requirements Engineering stage, which will be described in depth in the following sections.

The remainder of this paper is structured as follows. Section 2 presents background knowledge about Model-driven Architecture and the Software Process Engineering Metamodel. The proposal for a Requirements Engineering Method for Ubiquitous Systems (REUBI) is described in section 3. Then, in section 4, existing approaches are revised and compared to the proposal. Finally, in section 5, the conclusions of this work are summarized, as well as the future work.

2 Background

This section introduces the basics about the Model-Driven Development, a new paradigm which highlights clear benefits in the software development process. Related to it, the Software Process Engineering Metamodel will be used to define and support a software development process to design Ubiquitous and AmI Systems following a Model-Driven strategy.

2.1 Model-Driven Architecture (MDA)

Model-Driven Architecture (MDA) is an approach to the development, integration and interoperability of IT systems. The general strategy is to separate the specification of system operation from any technical consideration [10]. In particular, MDA defines three viewpoints:

- **Computation Independent Viewpoint (CIV):** focuses on the system environment and its requirements.
- **Platform Independent Viewpoint (PIV):** focuses on system operation, hiding the necessary details for a concrete platform.
- **Platform Specific Viewpoint (PSV):** focuses on the use of a specific platform by a system.

These viewpoints result in three system views at different levels of abstraction: the **Computation Independent Model** (CIM), the **Platform Independent Model** (PIM) and the **Platform Specific Model** (PSM), respectively.

The MDA lifecycle is similar to a traditional lifecycle [7]: requirements, analysis, design, coding, testing and deployment. However, one of the major differences lies in the nature of the artifacts that are created during the process as output of a phase and input for the following one. These artifacts are semiformal models, i.e. models that can be understood buy computers. The concept of model transformation is capital for obtaining the benefits of this lifecycle, such as portability and maintainability. As defined by OMG [10], transformation is *the process of converting one model to another model of the same system* (e.g. PIM to PSM). This process needs the source model to be marked according to a mapping (specifications to transformation) which is defined according to the target model [9]. The marked model can then be translated into the target model. This procedure is specially necessary for vertical transformations, where there is a different level of abstraction for each model [4]. In other words, the source model, at a higher level of abstraction, must be refined with additional information that is meaningful in the mode specific target model.

2.2 Software Process Engineering Metamodel (SPEM)

The **Software Process Engineering Metamodel** (SPEM), as proposed by OMG [11], aims to be able to define and describe any software development process and its components. It is based on the following concepts: an **Activity** is a set of work units which are coordinated in order to produce products or artifacts; a **Guidance** is a description about how activities should be done in order to effectively produce the products; **Roles** are resources which perform the activities and produce the **Products**, artifacts which are consumed and produced by the roles which perform the activities; finally, a **Phase** consists of significant segments of a complete software lifecycle, usually delimited by pre- and post-conditions, milestones and deliverables.

SPEM has been devised as a metamodel, but also specified in terms of a UML Profile, organized in different packages which extends the Core elements of the UML Specification [12]. This UML Profile will be used in the following sections in order to describe the proposal.

3 A Model-Driven Approach to Requirements Engineering in Ubiquitous Systems

Taking into account the concepts presented above, in this section the proposal, consisting of a Model-driven approach to the design of high quality Ubiquitous Systems (MD-UBI), is presented. As depicted in figure 1 (a), the MD-UBI process consists of four phases, inspired by the Unified Process [8]: **Requirements Engineering**, following a method named REUBI; **Design**, applying a method called SCUBI; **Implementation**; and **Testing**. In this paper, the focus is on the first stage, Requirements Engineering.

The Requirements Engineering phase distinguishes five different roles which take part in the different activities that are carried out. These roles, as shown in figure 1 (b), are: **Analyst, User**, **Stakeholder**, **Architect** and **Designer**. Whereas the last four roles only serve as assistants to the activities that need to be performed during this phase, the System Analyst is the main responsible of the success of this stage.

The REUBI process consists of several activities which have been formally defined. The models that are produced as a result of the execution of the different stages conform to the UML standard and have been specified in terms of metamodels. Several UML profiles have been defined in order to represent the concepts which are treated in every process step.

The activities the Analyst has to accomplish according to the REUBI method are (figure 1 (c)):

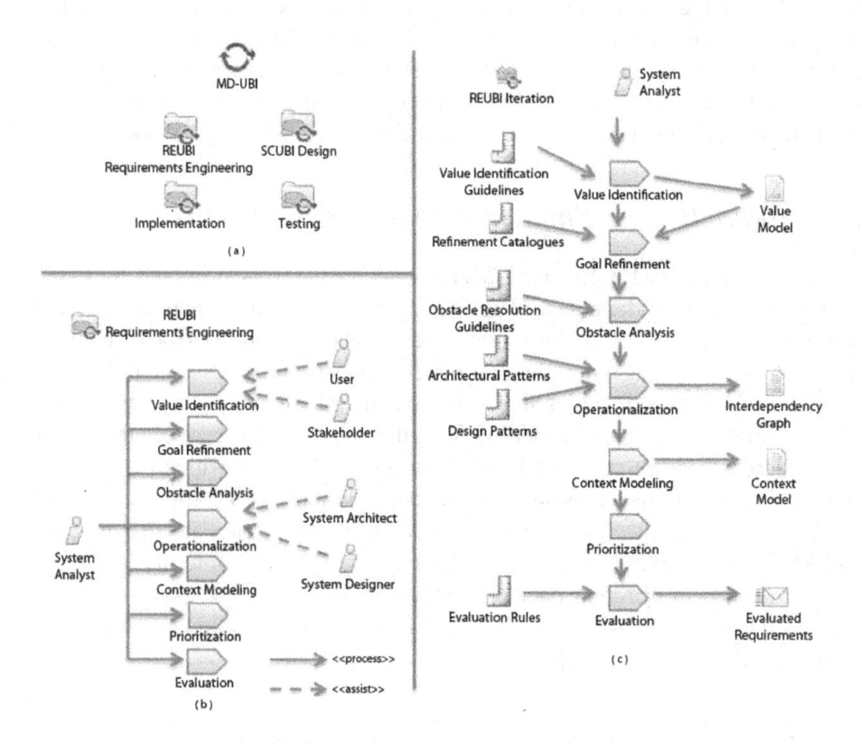

Fig. 1 SPEM Diagrams describing the proposal. (a) Package diagram showing the different phases in this process. (b) Use case diagram depicting the activities involved in the Requirements Engineering phase and the roles involved. (c) Sequence diagram describing one iteration of the Requirements Engineering stage.

- **Value identification:** the process is initiated by an initial discovery of the global requirements of the system. In order to achieve this, the System Analyst determines the *actors* which are relevant to the system and the values they exchange. A *value* is a good, service or piece of information which can be exchanged between two actors. Also, there are some aspects which can improve the quality of the value exchange; these are called *value enhancers*.

  This stage can make use of a set of guidelines that, posing appropriated questions, lead to the identification of actors, values and value enhancers. All this information is gathered into the *Value Model*.

- **Goal and Softgoal Refinement:** drawing from the *value model* defined in the previous step, the System Analyst needs to determine the top level *goals* and *softgoals* which must be fulfilled in the system-to-be. A goal is an objective which has a clear-cut criterion to determine its satisfaction. In contrast, softgoals lack this criterion; therefore, they are *satisficed* [3] when there is enough positive evidence to make this statement. As a rule of thumb, values are mapped to goals and value enhancers are mapped to softgoals.

  The concept of softgoal is particularly useful to represent NFRs. This kind of requirements are often related to quality properties which are difficult to be expressed without any ambiguity. Rather, some level of acceptance in its satisfaction can be determined, leading to the appearance of a softgoal.

  Top level goals and softgoals are decomposed into more specific ones until concrete architectural or design decisions to fulfill them can be found, representing them into an *Interdependency Graph*. For this task, the System Analyst can make use of *Decomposition Patterns*, reusing the knowledge which can be generated in other developments.
- **Obstacle Analysis:** ubiquitous computing environments are highly dynamical and heterogeneous; therefore, many inconvenient situations may arise. In order to be able to engineer robust, fault-tolerant ubiquitous systems, in this stage the System Analyst has to perform *obstacle analysis*. Sources of likely failure of the system need to be identified and related to the objectives they may have an impact to. Since the satisfaction of these objectives may be compromised, the System Analyst has to find solutions to the presence of obstacles. This leads to an iterative process of discovering of new obstacles and the corresponding objectives to mitigate their effects. The output of this stage is an *Interdependency Graph* which has been refined and augmented from the previous step.
- **Operationalization:** once the objectives have been decomposed, they need to be *operationalized*. An operationalization is a possible *architectural* or *design decision* which has a contribution to the satisfaction of the objective. Operationalizations can partially fulfill an objective, but also they may have side effects on the satisfaction of other objectives. Thus, negative relationships must also be reflected in this analysis.

  The results of decomposing and operationalizing the objectives are represented hierarchically and gathered into the *Interdependency Graph*. System Analysts can make use of catalogs of architectural or design patterns in order to apply well know solutions to the satisfaction of certain requirements.
- **Context Modelling:** an important feature of Ubiquitous and AmI systems is *context-awareness*. Systems must adapt their content, behavior or structure depending on context events. Thus, the impact of context on the requirements is studied. Rather than looking for the actual representation of context in the system-to-be, we focus on identifying the situations which are interesting for the system being engineered and the attributes which characterize it. This information is reflected in the *Context Model* and related to the *Interdependency Graph* to describe how context impacts on the requirements.

- **Prioritization:** unlike other kinds of systems, where requirements have a fixed priority during the execution of the system, ubiquitous systems requirements have a variable priority depending on changes in the environment. In this stage, the System Analyst must determine the priority each requirement receives according to the context situations identified in the previous stage. Three prioritization levels are distinguished: *critical* (indispensable for the success of the project), *important* (desirable for a high quality system) and *normal* (subject of trade-off if necessary).
- **Evaluation:** finally, the System Analyst need to perform an evaluation of the satisfaction of the requirements. For each different context situation involving a change of priorities in the requirements, the evaluation procedure must be applied. The System Analyst has to choose a set of operationalizations and, taking into account a set of *Evaluation Rules*, try to determine the satisfaction of the requirements. If any of the requirements is not satisfied, the process must be reapplied, trying to find new solutions until all the requirements are satisfied. As output of this stage, the System Analyst obtains a model with a decomposition of the requirements, their priorities, their operationalizations and the context situations which are relevant for the system, which can be delivered to System Architects and Designers in order to proceed with the design and implementation of the system.

REUBI is an adaptable and systematic process. It includes different steps which are complementary to each other. The employment of formal models to define what is done in every activity and to describe the deliverables guarantees the final results have a high quality. Thus, modifications in the models can be introduced iteratively at different levels and propagated through the different stages involved, as the Model-Driven approach suggests.

4 Related Work

Many approaches exist in the field of Requirements Engineering, some of them specially devised for Ubiquitous and AmI Systems. Although they take into account some of the above mentioned characteristics relative to the UC and AmI environments (e.g. context-awareness [14], personalization [15] or dinamicity [5]), they present some drawbacks, which have been addressed by the presented proposal:

- Most approaches do not use standard notations [1, 2], making the understanding and learning of the generated output more difficult. REUBI relies on the UML standard, both for its specification and for the construction of diagrams.
- Existing methods usually do not provide a systematic treatment of NFRs. The NFR Framework [3] presents a method to deal with NFRs; however, it is a general-purpose framework and does not take into account specific features of UC and AmI systems, as they are treated in this proposal.

- Some methods focus on the capture of requirements [6], whereas some others do on the requirements elaboration [14]. REUBI aims to support the complete Requirements Engineering stage.
- In a broader sense, existing techniques and frameworks take into account single and specific characteristics of UC and AmI Systems [14, 15]. The described proposal has been created taking into account additional features, although it still needs to be applied to engineering more ubiquitous systems in order to check its suitability.
- Existing Model-Driven approaches mainly focus on the design stage and briefly mentions the Requirements Engineering stage [13]. This proposal aims to provide a whole method which comprises the whole software lifecycle.

5 Conclusions and Future Work

Model-Driven Development guidelines provide numerous benefits such as easy maintenance, high productivity and high quality. Trying to take advantage of these features, in this paper we have introduced a Model-Driven method to develop Ubiquitous Systems, describing in depth the Requirements Engineering stage.

The proposed method conforms to the SPEM metamodel, which allows it to be easily included into existing CASE tools. Moreover, the notation to create the required models is based on UML, which eases its learning. REUBI has been applied to several case studies and real developments, demonstrating its feasibility to engineer the requirements of an Ubiquitous System and leading to high quality designs. The main benefits of this contribution are:

- It performs a systematic treatment of Non-Functional Requirements, which leads to a better understanding of them and consequently to a high quality design. This has been achieved by the formal definition of both the process and the models, and thanks to the existence of an evaluation process which helps the Analyst to determine which are the best options to be taken in order to satisfy the expected quality properties.
- It takes into account the special characteristics of Ubiquitous Systems, such as dinamicity, context-awareness or adaptivity, providing a further analysis and elaboration of the requirements that need to be fulfilled under different circumstances.
- The proposed models conform to the UML standard in order to apply a model-driven strategy and derive semi-automatically high quality designs from the requirements analysis carried out by means of the REUBI approach.

As future work, we plan to apply the proposed method to a wide variety of ubiquitous systems in order to gain knowledge about its strengths and weaknesses in its applicability. Moreover, the usefulness of the deliverables to derive high quality designs need to be determined, and the proposed models may need to be adjusted in order to deal with the special features of Ubiquitous Systems in a better way.

Besides, we plan to continue working on the other stages of this Model-driven approach in order to provide a complete methodology, and associated development tools based on standards, to engineer Ubiquitous Systems from the early stages of conception to a fully implemented working system.

Acknowledgements. The Spanish Ministry of Science and Innovation funds this research work through the projects TIN2008-05995/TSI.

References

1. Baresi, L., Pasquale, L., Spoletini, P.: Fuzzy Goals for Requirements-Driven Adaptation. In: IEEE International Conference on Requirements Engineering, pp. 125–134 (2002)
2. Cheng, B.H.C., Sawyer, P., Bencomo, N., Whittle, J.: A Goal-Based Modeling Approach to Develop Requirements of an Adaptive System with Environmental Uncertainty. In: Schürr, A., Selic, B. (eds.) MODELS 2009. LNCS, vol. 5795, pp. 468–483. Springer, Heidelberg (2009)
3. Chung, L., Nixon, B.A., Yu, E., Mylopoulos, J.: Non-Functional Requirements in Software Engineering. Kluwer Academic Publishers (2000)
4. De Lara, J., Vangheluwe, H.: Defining Visual Notations and Their Manipulation Through Meta-Modelling and Graph Transformation. Journal of Visual Languages and Computing. Special issue on Domain-Specific Modeling with Visual Languages 15(3-4), 309–330 (2004)
5. Goldsby, H.J., Sawyer, P., Bencomo, N., Cheng, B.H.C., Hughes, D.: Goal-based modeling of dynamically adaptive system requirements. In: IEEE International Conference on the Engineering of Computer-Based Systems, pp. 36–45 (2008)
6. Jrgensen, J.B., Bossen, C.: Executable use cases: Requirements for a pervasive health care system. IEEE Softw. 21, 34–41 (2004)
7. Kleppe, A., Warmer, J., Bast, W.: MDA Explained - The Model Driven Architecture: Practice and Promise. Addison Wesley (2003)
8. Kruchten, P.: The Rational Unified Process: an introduction. Addison-Wesley (2004)
9. Lankhorst, M.: Enterprise architecture modelling. The issue of integration. Advanced Engineering Informatics 18(4), 205–216 (2004)
10. OMG: MDA Guide. Version 1.0.1. In: Miller J., Mukerji J. (eds.) OMG (June 2003), `http://www.omg.org/cgi-bin/doc?omg/03-06-01`
11. OMG: Software and Systems Process Engineering Metamodel Specification. OMG (April 2008), `http://www.omg.org/spec/SPEM/2.0/`
12. OMG: Unified Modelling Language Specification. OMG (November 2007), `http://www.omg.org/spec/UML/2.1.2/`
13. Serral, E., Valderas, P., Pelechano, V.: Towards the Model Driven Development of context-aware pervasive systems. Pervasive and Mobile Computing, 254–280 (2010)
14. Sitou, W., Spanfelner, B.: Towards requirements engineering for context adaptive systems. In: Annual International Conference on Computer Software and Applications, pp. 593–600 (2007)
15. Sutcliffe, A., Fickas, S., Sohlberg, M.M.: Personal and contextual requirements engineering. In: IEEE International Conference on Requirements Engineering, pp. 19–30 (2005)

Optimizing Dialog Strategies for Conversational Agents Interacting in AmI Environments

David Griol, Javier Carbó, and José Manuel Molina

Abstract. In this paper, we describe a conversational agent which provides academic information. The dialog model of this agent has been developed by means of a statistical methodology that automatically explores the dialog space and allows learning new enhanced dialog strategies from a dialog corpus. A dialog simulation technique has been applied to acquire data required to train the dialog model and then explore the new dialog strategies. A set of measures has also been defined to evaluate the dialog strategy. The results of the evaluation show how the dialog model deviates from the initially predefined strategy, allowing the conversational agent to tackle new situations and generate new coherent answers for the situations already present in the initial corpus. The proposed technique can be used not only to develop new dialog managers but also to explore new enhanced dialog strategies focused on user adaptation required to interact in AmI environments.

Keywords: Conversational Agents, Speech Interaction, Agent & Multiagent Systems for AmI, Statistical Methodologies.

1 Introduction

Ambient Intelligence (AmI) and Smart Environments (SmE) emphasize on greater user-friendliness, more efficient services support, user-empowerment, and support for human interactions. For this reason, AmI systems usually consist of a set of interconnected computing and sensing devices which surround the user pervasively in his environment and are invisible to him, providing a service that is dynamically adapted to the interaction context, so that users can naturally interact with the system and thus perceive it as intelligent.

David Griol · Javier Carbó · José Manuel Molina
Group of Applied Artificial Intelligence (GIAA), Computer Science Department, Carlos III University of Madrid
e-mail: {david.griol, javier.carbo, josemanuel.molina}@uc3m.es

P. Novais et al. (Eds.): Ambient Intelligence - Software and Applications, AISC 153, pp. 93–100.
springerlink.com

To ensure such a natural and intelligent interaction, it is necessary to provide an effective, easy, safe and transparent interaction between the user and the system. With this objective, as an attempt to enhance and ease human-to-computer interaction, in the last years there has been an increasing interest in simulating human-to-human communication, employing conversational agents [4].

A conversational agent can be defined as a software that accepts natural language as input and generates natural language as output, engaging in a conversation with the user. In a conversational agent of this kind, several modules cooperate to perform the interaction with the user: the Automatic Speech Recognizer (ASR), the Language Understanding Module (NLU), the Dialog Manager (DM), the Natural Language Generation module (NLG), and the Synthesizer (TTS). Each one of them has its own characteristics and the selection of the most convenient model varies depending on certain factors: the goal of each module, or the capability of automatically obtaining models from training samples.

The application of statistical approaches to dialog management has attracted increasing interest during the last decade [8]. Statistical models can be trained from real dialogs, modeling the variability in user behaviors. The final objective is to develop conversational agents that have a more robust behavior and are easier to adapt to different user profiles or tasks. The most extended methodology for machine-learning of dialog strategies consists of modeling human-computer interaction as an optimization problem using Partially Observable Markov Decision Processes (POMDPs) and reinforcement methods. However, they are limited to small-scale problems, since the state space would be huge and exact POMDP optimization would be intractable [7].

In addition, the success of these approaches depends on the quality of the data used to develop the dialog model. Considerable effort is necessary to acquire and label a corpus with the data necessary to train a good model. A technique that has currently attracted an increasing interest is based on the automatic generation of dialogs between the dialog manager and an additional module, called the user simulator, which represents user interactions with the conversational agent [6]. The construction of user models based on statistical methods has provided interesting and well-founded results in recent years and is currently a growing research area. Therefore, these models can be used to learn a dialog strategy by means of its interaction with the conversational agent and reduce the effort to acquire a dialog corpus.

In this paper, we present a technique for learning optimal dialog strategies in conversational agents. Our technique is based on the use of a statistical dialog manager that is learned using a dialog corpus for the specific task. A dialog simulation technique is used to automatically generate the data required to learn a new dialog model. We have applied our technique to explore dialog strategies for a conversational agent designed to provide academic information. In addition, a set of specific measures has been defined to evaluate the new strategy once new simulated data is used to re-train the dialog manager. The results of the evaluation of a dialog manager developed for this agent show how the variability of the dialog model is increased by detecting new dialog situations that are not present in an initial model and selecting better system responses for the situations that were already present.

2 Our Statistical Dialog Management Technique

In most conversational agents, the conversational agent takes its decisions based only on the information provided by the user in the previous turns and its own model. This is the case with most slot-filling dialogs. The methodology that we propose for the selection of the next system answer in this kind of task is as follows [2]. We consider that, at time *i*, the objective of the dialog manager is to find the best system answer A_i. This selection is a local process for each time *i* and takes into account the previous history of the dialog, that is to say, the sequence of states of the dialog (i.e. pairs *system-turn, user-turn*) preceding time *i*:

$$\hat{A}_i = \underset{A_i \in \mathscr{A}}{\operatorname{argmax}} P(A_i|S_1, \cdots, S_{i-1})$$

where set $\mathscr{A}$ contains all the possible system answers.

As the number of all possible sequences of states is very large, we define a data structure in order to establish a partition in the space of sequences of states (i.e., in the history of the dialog preceding time *i*). This data structure, that we call Dialog Register (*DR*), contains the information provided by the user throughout the previous history of the dialog. The selection of the best A_i is then given by:

$$\hat{A}_i = \underset{A_i \in \mathscr{A}}{\operatorname{argmax}} P(A_i|DR_{i-1}, S_{i-1})$$

The selection of the system answer is carried out through a classification process, for which a multilayer perceptron (MLP) is used. The input layer receives the codification of the pair (DR_{i-1}, S_{i-1}). The output generated by the MLP can be seen as the probability of selecting each of the different system answers defined for a specific task.

3 Our Dialog Simulation Technique

Our approach for acquiring a dialog corpus is based on the interaction of a user simulator and a conversational agent simulator [3]. Both modules use a random selection of one of the possible answers defined for the semantics of the task (user and system dialog acts). At the beginning of the simulation, the set of system answers is defined as equiprobable. When a successful dialog is simulated, the probabilities of the answers selected by the dialog manager during that dialog are incremented before beginning a new simulation.

An error simulator module has been designed to perform error generation. The error simulator modifies the frames generated by the user simulator once it selects the information to be provided. In addition, the error simulator adds a confidence score to each concept and attribute in the frames. The model employed for introducing errors and confidence scores is inspired in the one presented in [5]. Both processes are carried out separately following the noisy communication channel metaphor by means of a generative probabilistic model $P(c, a_u|\tilde{a}_u)$, where a_u is the true incoming user dialog act $\tilde{a}_u$ is the recognized hypothesis, and c is the confidence score associated with this hypothesis.

On the one hand, the probability $P(\tilde{a}_u|a_u)$ is obtained by Maximum-Likelihood using the initial labeled corpus acquired with real users. To compute it, we consider the recognized sequence of words w_u and the actual sequence uttered by the user $\tilde{w}_u$.

$$P(\tilde{a}_u|a_u) = \sum_{\tilde{w}_u} P(a_u|\tilde{w}_u) \sum_{w_u} P(\tilde{w}_u|w_u) P(w_u|a_u)$$

On the other hand, the generation of confidence scores is carried out by approximating $P(c|\tilde{a}_u, a_u)$ assuming that there are two distributions for c. These two distributions are defined manually generating confidence scores for correct and incorrect hypotheses.

$$P(c|a_w, \tilde{a}_u) = \begin{cases} P_{corr}(c) & if \ \ \tilde{a}_u = a_u \\ P_{incorr}(c) & if \ \ \tilde{a}_u \neq a_u \end{cases}$$

4 Design of an Academic Conversational Agent

The design of our conversational agent is based on the requirements defined for a dialog system developed to provide spoken access to academic information about the Department of Languages and Computer Systems in the University of Granada [1]. To successfully manage the interaction with the users, the conversational agent carries out six main tasks described in the Introduction section: automatic speech recognition (ASR), natural language understanding (NLU), dialog management (DM), database access and storage (DB), natural language generation (NLG), and text-to-speech synthesis (TTS). The information that the conversational agent provides has been classified in four main groups: subjects, professors, doctoral studies and registration.

The semantic representation that we have chosen for the task is based on the concept of frame, in which one or more concepts represent the intention of the utterance, and a sequence of attribute-value pairs contains the information about the values given by the user. In the case of user turns, we defined four concepts related to the different queries that the user can perform to the system (*Subject*, *Lecturers*, *Doctoral studies*, and *Registration*), three task-independent concepts (*Affirmation*, *Negation*, and *Not-Understood*), and eight attributes (*Subject-Name*, *Degree*, *Group-Name*, *Subject-Type*, *Lecturer-Name*, *Deadline,Program-Name*, *Semester*, and *Deadline*).

The labeling of the system turns is similar to the labeling defined for the user turns. A total of 30 task-dependent concepts was defined:

- Task-independent concepts (*Affirmation*, *Negation*, *Not-Understood*, *New-Query*, *Opening*, and *Closing*).
- Concepts used to inform the user about the result of a specific query (*Subject*, *Lecturers*, *Doctoral-Studies*, and *Registration*).

- Concepts defined to require the user the attributes that are necessary for a specific query (*Subject-Name*, *Degree*, *Group-Name*, *Subject-Type*, *Lecturer-Name*, *Program-Name*, *Semester*, and *Deadline*).
- Concepts used for the confirmation of concepts (*Confirmation-Subject*, *Confirmation-Lecturers*, *Confirmation-DoctoralStudies*, *Confirmation-Registration*) and attributes (*Confirmation-SubjectName*, *Confirmation-Degree*, *Confirmation-SubjectType*, *Confirmation-LecturerName*, *Confirmation-Semester*, *Confirmation--ProgramName*, *Confirmation-Deadline*, and *Confirmation-GroupName*).

The *DR* defined for the task is a sequence of 12 fields, corresponding to the four concepts (*Subject*, *Lecturers*, *Doctoral-Studies*, and *Registration*) and eight attributes (*Subject-Name*, *Degree*, *Group-Name*, *Subject-Type*, *Lecturer-Name*, *Program-Name*, *Semester*, and *Deadline*) defined for the task.

A set of 100 dialogs was acquired with the conversational agent by means of its interaction with real users using an initial version of the system [1]. The acquisition process resulted in a spontaneous Spanish speech dialog corpus with 60 different speakers. The total number of user turns was 422. The recorded material has a duration of 150 minutes. Figure 1 shows an example of a dialog extracted from this corpus. Turns with S refer to system turns, and turns with U refer to user turns.

S1: Welcome to the university information system. You can consult information about subjects, lecturers, doctoral studies and registrations.
U1: I want to know information about subjects in Computer Science.
S2: Which subject do you want to consult?
U2: Language Processors I.
S3: Here you can consult the information of the subject. Anything else?
U3: Can you inform me about the schedules of Group D?
S4: I give the information for the Group D of Language Processors I: Monday 9:00-11:00, Wednesday 11:00-13:00. Anything else?
U4: Can you tell me the name of the lecturer?
S5: I give the information for the Group D of Language Processors I. The name of the lecturer is John Smith. Anything else?
U5: No, thank you.
S6: Thank you for using the system.

Fig. 1 Example of a dialog acquired with the interaction of the initial conversational agent with real users

5 Evaluation of Our Proposal

A dialog corpus of 3000 successful dialogs was acquired using the proposed user simulation technique following the same objectives defined for the initial acquisition with real users. A maximum number of 14 user turns per dialog was defined for the acquisition.·

We have considered different dialog style features to evaluate the initial conversational agent for the task and its evolution once the simulated dialogs are incorporated to learn a new dialog model for the conversational agent. We defined and counted a set of system/user dialog acts. On the system side, we have measured the confirmation of concepts and attributes, questions to require information, and system answers generated after a database query. On the user side, we have measured the percentage of turns in which the user carries out a request to the system, provides information, confirms a concept or attribute, the Yes/No answers, and other answers not included in the previous categories. Finally, we have measured the proportion of goal-directed actions (request and provide information) versus the grounding actions (confirmations) and rest of actions.

Table 1 show the frequency of the most dominant user and system dialog acts in the initial and final conversational agents. From its comparison, it can be observed that there are significant differences in the dialog acts distribution. With regard to user actions, it can be observed that users need to employ less confirmation turns in the final agent, which explains the higher proportion for the rest of user actions using the final conversational agent. It also explains the lower proportion of yes/no actions in the final agent, which are mainly used to confirm that the system's query has been correctly provided. With regard to the system actions, it can be observed a reduction in the number of system confirmations for data items. This explains a higher proportion of turns to inform and provide data items for the final agent. Both results show that the final conversational agent carries out a better selection of the system responses.

Table 1 Percentages of different types of user dialog acts (top) and system dialog acts (bottom)

	Initial Conversational Agent	Final Conversational Agent
Request to the system	31.74%	35.43%
Provide information	20.72%	24.98%
Confirmation	10.81%	7.34%
Yes/No answers	31.47%	28.77%
Other answers	3.26%	3.48%

	Initial Conversational Agent	Final Conversational Agent
Confirmation	13.51%	10.23%
Questions to require information	18.44%	19.57%
Answers after a database query	68.05%	70.20%

In addition, we grouped all user and system actions into three categories: "goal directed" (actions to provide or request information), "grounding" (confirmations and negations), and "rest". Table 2 shows a comparison between these categories. As can be observed, the dialogs provided by the final conversational agent have a better quality, as the proportion of goal-directed actions is higher.

Finally, a total of 100 dialogs was recorded from interactions of 10 students and professors of our University employing the conversational agent developed using our proposal. We considered the following measures for the evaluation:

- Dialog success rate (*%success*). Percentage of successfully completed tasks;
- Average number of user turns per dialog (*nT*);
- Confirmation rate (*%confirm*). Ratio between the number of explicit confirmations turns (nCT) and the number of turns in the dialog (nCT/nT);
- Average number of corrected errors per dialog (*nCE*). Average of errors detected and corrected by the dialog manager;
- Average number of uncorrected errors per dialog (*nNCE*). Average of errors not corrected by the dialog manager;
- Error correction rate (*%ECR*). Percentage of corrected errors, computed as nCE/ (nCE + nNCE).

Table 2 Proportions of dialog spent on-goal directed actions, ground actions and the rest of possible actions

	Initial Conversational Agent	Final Conversational Agent
Goal directed actions	62.55%	70.17%
Grounding actions	36.32%	29.59%
Rest of actions	1.13%	1.24%

Table 3 compares this acquisition with the acquisition of 100 dialogs with real users using the initial conversational agent. The results show that both conversational agents could interact correctly with the users in most cases. However, the final conversational agent system obtained a higher success rate, improving the initial results by 6% absolute. Using the final conversational agent, the average number of required turns is also reduced from 4.99 to 3.75. The confirmation and error correction rates were also improved by the final conversational agent, as the enhanced dialog model reduces the probability of introducing ASR errors. The main problem detected was related to the introduction of data in the *DR* with a high confidence value due to errors generated by the ASR not detected by the dialog manager.

Table 3 Results of the evaluation of the conversational agents with real users

	%success	nT	%confirm	nCE	nNCE	%ECR
Initial Conversational Agent	89%	4.99	34%	0.84	0.18	82%
Final Conversational Agent	95%	3.75	37%	0.89	0.07	92%

6 Conclusions

In this paper, we have described a technique for exploring dialog strategies in conversational agents. Our technique is based on two main elements: a statistical dialog methodology for dialog management and an automatic dialog simulation technique to generate the data that is required to re-train the dialog model. The results of applying our technique to the design of conversational agent that provides academic information show that the proposed methodology can be used not only to develop new dialog managers but also to explore new enhanced strategies required for the interaction in AmI environments. Carrying out these tasks with a non-statistical approach would require a very high cost that sometimes is not affordable. As a future work, we are adapting the proposed dialog management to evaluate the capability of our methodology to adapt efficiently to AmI environments that vary dynamically, in which additional information sources (including context information) must be considered in the definition of the *DR* and additional modalities and more complex operations are required for the interaction with users.

Acknowledgements. Research funded by projects CICYT TIN2011-28620-C02-01, CICYT TEC2011-28626-C02-02, CAM CONTEXTS (S2009/TIC-1485), and DPS2008-07029-C02-02.

References

1. Callejas, Z., López-Cózar, R.: Relations between de-facto criteria in the evaluation of a spoken dialogue system. Speech Communication 50(8–9), 646–665 (2008)
2. Griol, D., Hurtado, L., Segarra, E., Sanchis, E.: A Statistical Approach to Spoken Dialog Systems Design and Evaluation. Speech Communication 50(8-9), 666–682 (2008)
3. Griol, D., Sánchez-Pi, N., Carbó, J., Molina, J.: An Agent-Based Dialog Simulation Technique to Develop and Evaluate Conversational Agents. In: Proc. of PAAMS 2011. AISC, vol. 88, pp. 255–264 (2011)
4. López-Cózar, R., Araki, M.: Spoken, Multilingual and Multimodal Dialogue Systems. John Wiley & Sons Publishers (2005)
5. Schatzmann, J., Thomson, B., Weilhammer, K., Ye, H., Young, S.: Agenda-Based User Simulation for Bootstrapping a POMDP Dialogue System. In: Proc. of HLT/NAACL 2007, pp. 149–152 (2007)
6. Schatzmann, J., Weilhammer, K., Stuttle, M., Young, S.: A Survey of Statistical User Simulation Techniques for Reinforcement-Learning of Dialogue Management Strategies. Knowledge Engineering Review 21(2), 97–126 (2006)
7. Williams, J., Young, S.: Scaling POMDPs for Spoken Dialog Management. IEEE Audio, Speech and Language Processing 15(8), 2116–2129 (2007)
8. Young, S.: The Statistical Approach to the Design of Spoken Dialogue Systems. Tech. rep., Cambridge University Engineering Department (UK) (2002)

Analysing Participants' Performance in Idea Generation Meeting Considering Emotional Context-Aware

João Carneiro, João Laranjeira, Goreti Marreiros, and Paulo Novais

Abstract. In an idea generation meeting the emotional context-aware is correlated the with participants' performance. To achieve good results in idea generation sessions it is important that participants maintain a positive emotional state. In this paper we will present an agent-based idea generation meeting using an emotional context-aware model that includes several mechanisms, namely: the analysis and evaluation of events, emotion-based process and recommendation selection. In order to analyze and test the benefits of these mechanisms, we also created a simulated scenario in which agents generate ideas and are affected by context-aware situations. The simulation reveals that these mechanisms can increase the participants' performance using some techniques. This study shows how the use of recommendation mechanisms can maximize meeting results in some situations.

Keywords: emotional context-aware, idea generation meeting, agent based simulation.

Introduction

In the last years the role of emotions is more correlated with several cognitive activities, namely: decision-making, learning, planning and creative problem solving [1][2]. The idea generation process, a part of the creative problem solving

---

João Carneiro · João Laranjeira · Goreti Marreiros
GECAD – Knowledge Engineering and Decision Support Group

Goreti Marreiros
Institute of Engineering – Polytechnic of Porto

Paulo Novais
University of Minho
e-mail: jmrc@isep.ipp.pt, jpcl@isep.ipp.pt, mgt@isep.ipp.pt, pjon@di.uminho.pt

P. Novais et al. (Eds.): Ambient Intelligence - Software and Applications, AISC 153, pp. 101–108.
springerlink.com

cycle, is ever more used by organizations to create new products or services, new strategies and to change organization structures. Since organizations are made by multiple people from different areas, usually idea generation is done in groups, in which interaction can be beneficial for the appearance of new ideas [3]. However, these interactions can affect negatively the participants' performance and consequently the group performance [4]. Based on the assumption that the emotions are triggered by events, it is possible to conclude that the individuals' mood will be affected by social interactions. This fact is unfavorable to group idea generation as various studies have proved [5][6][7].

According to Ortony [8], an emotion is the result of three types of subjective evaluations: evaluation of the triggered events considering the defined goals; evaluation of the actions taken by a certain agent, and also by the evaluation of the agent's own attitudes. This way, regarding the process of group idea generation, where events are triggered constantly, one can easily conclude that there is an emotional context associated to the process. According to the literature, it is possible to conclude that the emotional context of an idea generation meeting influences the performance of the participants. Several studies found in the literature prove that when the participants are in a positive mood, they generate more ideas and more creative ideas [5][6][9][10][11].

In this paper is used a model that aims to infer every motional context of the meeting, allowing, this way, all the participants to spend as much time as possible over the process in a positive mood. This fact allows ideas to be generated in bigger quantities and with more creativity. This type of implementation is useful for generating ideas of big groups or groups that are not in the same space.

In order to validate the correct emotional context inference by the model, and also the increase of the participants' productivity in the group idea generation process a set of experiments were conducted. The experiments consisted in a set of simulations based on agents (agent-based simulation), in which the participating agents that represented the members of the idea generation group were modeled with profiles to have a certain type of action.

In the rest of the paper we first present the emotional context of group idea generation and how to model it. Next we present an agent based architecture to simulate an idea generation meeting scenario and the emotional context of this meeting. For that we present some experiments to evaluate the model and understand what is the influence of emotional context in group idea generation. In the next section we analyze the results of experiments and in the last section we present the conclusions and future work.

Emotional Context Modeling

In order to understand and simulate the emotional context of the group idea generation process it is necessary to model the emotional aspects of each participant in particular. Knowing that the emotional context of an idea generation meeting varies according to the events that happen and knowing which is the influence of those events on the participants' mood is an essential task. This way, in order to constantly adapt the meetings emotional context, facilitating actions are taken aiming to maintain the participants in a positive mood. These actions contribute directly to the maximization of the participants' performance and, consequently, to the maximization of the idea generation meeting results.

The events considered in the emotional context modeling may include the introduction of new ideas, the evaluation of the ideas, the visualization and analysis of the performance and the reception of facilitation recommendations. The participants are emotionally affected by these events, with the possibility of being positively or negatively affected, according to the desirability the participant has for that event.

The emotional context model [12] presented in this section was developed on basis on 3 assumptions, resulting from the literature review:

1. The individuals have a tendency to have better performance (generate more ideas) when they are in a positive mood;
2. The ideas generated in a positive mood have the tendency to be more creative responses;
3. The inclusion of a facilitator in the group idea generation process has the goal of improving the group performance (generate more ideas);

This model is based on events, which are the input of the model, and applies the OCC model [8] to infer the participants' emotions. After calculating the desirability that each participant has for the event, one or more emotions are generated considering that same desirability. Since the mood represents the participants emotional state over the time, if the participant is in a negative mood, then recommendations are generated regarding the events that led the participant in that mood. This way, the output of the model is the recommendations which have the goal of keeping the participant in a positive mood, facilitating therefore the participant's performance.

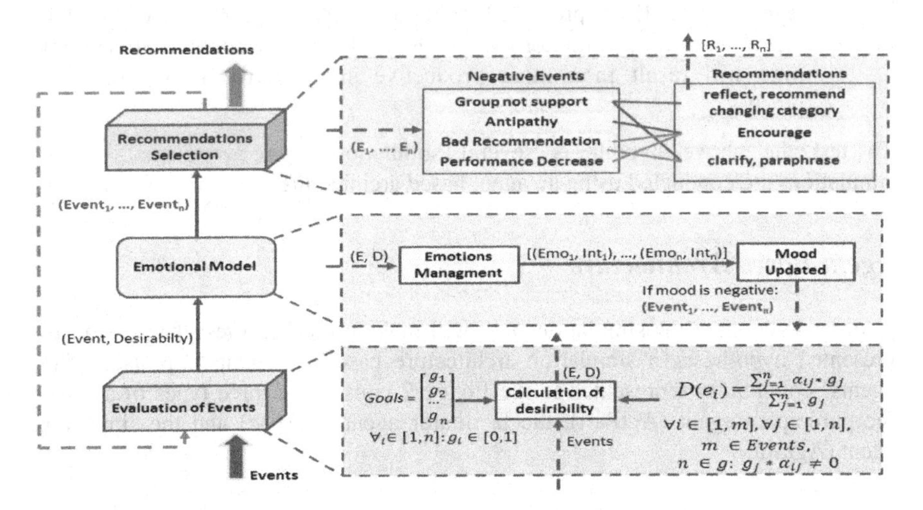

Fig. 1 Emotional Context Model [12]

The **Evaluation of Events** component has the goal of evaluating the impact that a certain event has in the participant, i. e., the participant's desirability for a triggered event. The desirability is calculated considering the impact that the event has in the participant's objectives and also considering the importance of each in the participant's performance.

The **Emotional Model** component infers that the emotions are generated in the participant. Considering the triggered events and the participant's desirability for its happening, certain emotions are generated that, in turn, will affect the participant's mood. Each emotion follows a certain rule that is triggered whenever an event happens. If the triggered event is a desired event, then a positive emotion is generated, otherwise a negative emotion is generated. The participant's mood is calculated by the sum of the positive emotions with the negative emotions, always considering that over the time the emotions' intensity will decline.

The **Recommendations Selection** component aims to generate recommendations that will be sent to the participant with a negative mood. As regards to the negative events that led the participants in that state, recommendations to "put" the participant in a positive mood are generated. However, the generated recommendations may not have a positive influence in the participant, because he/she can evaluate negatively the received recommendations.

Case Study

The ideas presented above were developed into hypotheses in order to test the model presented in the previous section. The hypotheses are the following:

- Hypothesis 1: The participants who are in a positive mood for a long period of time generate more ideas;
- Hypothesis 2: If the presented model analyses the events correctly, then the presence of a facilitator who sends good recommendations at the right time will result in a more productive and creative idea generation process.

To test the above hypothesis several simulations were conducted. These simulations are conducted using an agent based architecture.

Agent Based Architecture

In order to conduct the simulations that will make possible to test the previously presented hypothesis, a simulation architecture based on agents was used. The agents' community represented in the Figure 2 consists of three types of agents: the participant agent (AgtPart), the facilitator agent (AgtFac) and the simulator agent (AgtSim).

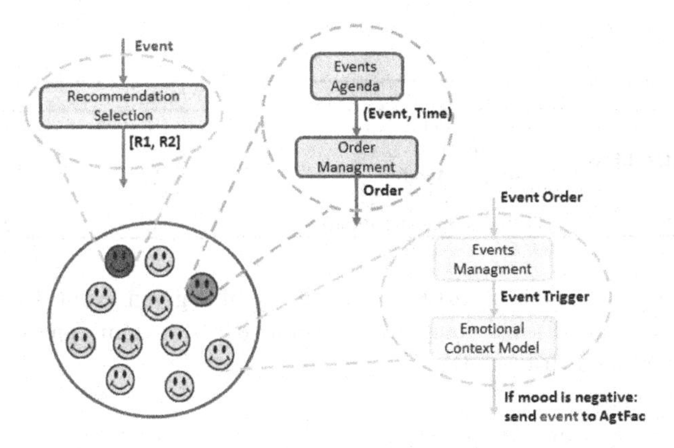

Fig. 2 Agent Based Architecture to Group Idea Generation Simulation

The **participant agent** represents the participant of idea generation meeting and has the capability to trigger the events scheduled by the AgtSim and also to infer the emotional context model. The **facilitator agent** represents the idea generation meeting facilitator and her function is to generate recommendations to send to one AgtPart, whenever he/she is in a negative mood. At last, the **agent simulator** is responsible for generating the entire simulation, i. e., schedule the events that will be triggered and send the execution order to the AgtPart that will trigger those events.

Simulations

30 simulations were conducted for each one of the three experiments described below where, to better understand the difference in the number of ideas generated, the time for each simulation is equivalent to 120 minutes. Experiment 1 was conducted without the facilitation process; Experiment 2 was conducted with a good facilitation process; and Experiment 3 was conducted with a bad facilitation process. The focus of each experience will be a participant agent that had the profile presented in Table 1.

Table 1 Participant Desirability

	Desirability
Group Support	0.68
Group not Support	-0.55
Performance Increase	0.75
Performance Decrease	-0.58

	Desirability
Good Recommendations	0.80
Bad Recommendations	-0.80
Sympathy of others	0.80
Antipathy of others	-0.80

Table 2 represents the average time ($\bar{x}$) of an event to be triggered and the standard deviation (σ) of this average time. These data allow generating random values based on a normal distribution.

Table 2 Average time to events triggered

		$\bar{x}$ (minutes)	σ
Generated Ideas	Positive Mood	4	2
	Neutral Mood	7	4
	Negative Mood	12	5
Analyse	Performance or Empathy	8	5

Table 3 represents the occurrence probability of a given event. Considering a probability average ($\bar{x}$) and standard deviation (σ), a random value is generated based on a normal distribution.

Table 3 Probability of events triggered

		$\bar{x}$ (Probability)	σ
Facilitation Quality	Good	0.7	0.2
	Bad	0.2	0.2
	Group Support	0.71	0.47
	Sympathy	0.56	0.4

Table 4 shows the number of ideas generated by the participating agents in each of the experiments.

Table 4 Idea Generated in three experiments

	Generated Ideas	Average of ideas (30 Simulations)	Standard Deviation
1st Experiment	565	18.83	4.14
2nd Experiment	664	22.13	4.56
3rd Experiment	511	17.03	4.09

In Table 5 are presented the results for the number of ideas generated in each experiment when the participant agent was in a positive, neutral or negative mood.

Table 5 Ideas Generated by participant by mood

	Positive Mood			Neutral Mood			Negative Mood		
	Ideas	Avg	StDev	Ideas	Avg	StDev	Ideas	Avg	StDev
1st Exp	388	12.93	5.03	90	3	1.58	87	2.9	1.49
2nd Exp	529	17.63	4.53	111	3.7	1.68	24	0.8	1.85
3rd Exp	319	10.63	4.50	78	2.6	1.85	114	3.8	1.67

Discussion

By analysing the results obtained in the experiments conducted, one can conclude that the inclusion of a good facilitator in the presented model, the participant agent generated a higher number of ideas when comparing to the experiment that had no facilitator and to the experiment that included a bad facilitator. One can also see in Figure 3 that the largest amount of ideas generated was verified in almost every simulation.

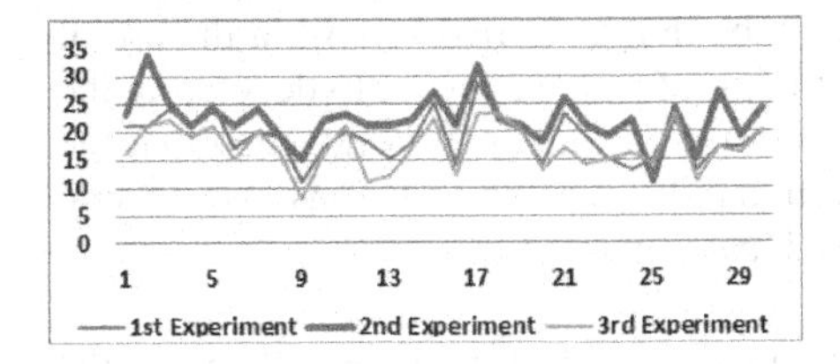

Fig. 3 Chart of idea generated in three experiments

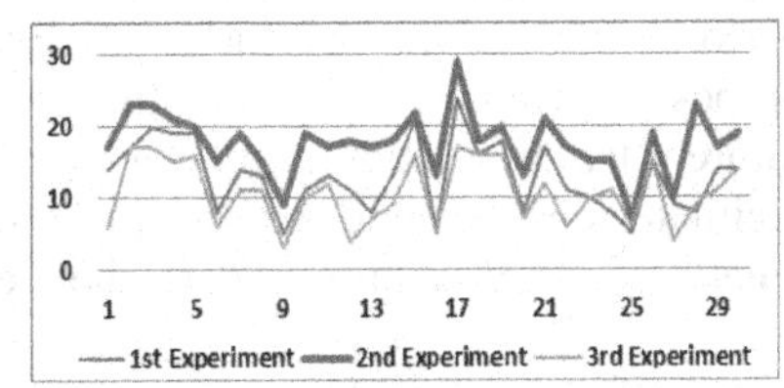

Fig. 4 Chart of Idea generated by participant in positive mood

Analysing Figure 4 we can conclude that when there is a good facilitation the participants generate more ideas in a positive mood. As the participants in positive mood generate more creative ideas [6] then we can conclude that the model presented [12], when is used a good facilitation, facilitates the generation of more creative ideas.

Figure 5 shows that there is no tendency in none of the experiments where are analyses of the data about number of idea generated in neutral mood.

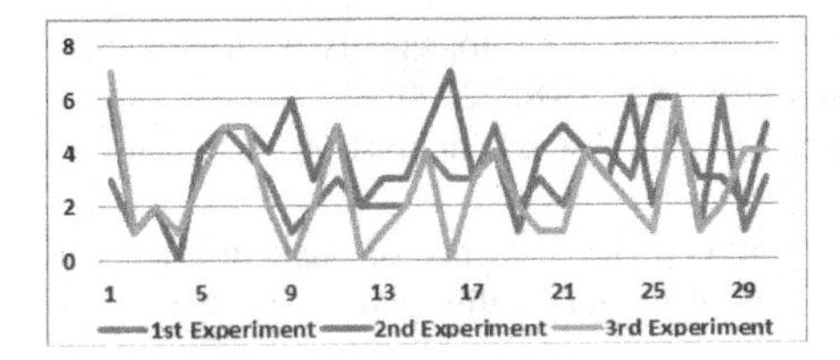

Fig. 5 Chart of Idea generated by participant in neutral mood

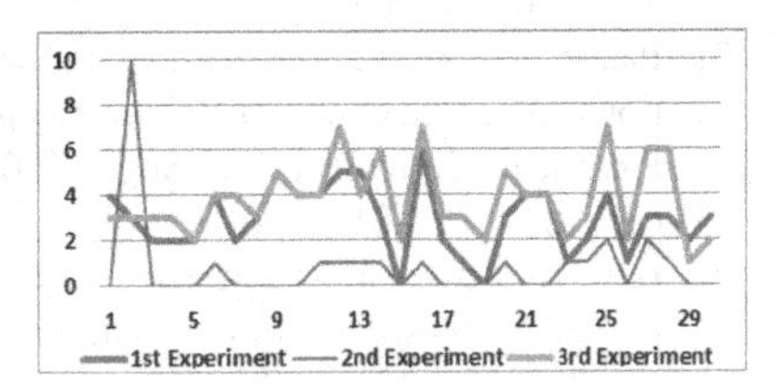

Fig. 6 Chart of Idea generated by participant in negative mood

Figure 6 shows how the number of ideas generated in a negative mood is much higher when there is not a facilitator or when there is a bad facilitator.

In conclusion, the model presented can be aware of the various actions that take place throughout the process that can affect the participant's mood. The inclusion of a good facilitator can increase significantly the number of ideas generated by the participant agent, as well as to increase the number of ideas generated in a positive mood and decrease the ideas generated in a negative mood. Thus, we can prove the hypothesis 1 and hypothesis 2.

Conclusions and Future Work

With the results obtained it is possible to see the importance of the inclusion of a good facilitator. It was possible to understand that the model represents correctly the events of an idea generation process and the facilitator has the correct acting timing. Obviously the improvements presented in the participants' performance are not just due to the presence of the facilitator, but to the way the facilitator takes part of the process. With the conducted experiments, one can verify that when recommendations sent by the facilitator are good, they improve the performance of the group members. Even though the verified benefit in the use of proposed recommendations and in their timing of use, we intend to develop in the future a learning module to be integrated in the facilitator. It is intended that the facilitator may become increasingly assertive in his recommendations over several meetings, as he/she knows more about each participant.

Acknowledgments. This work is supported by FEDER Funds through the "Programa Operacional Factores de Competitividade – COMPETE" program and by National Funds through FCT "Fundação para a Ciência e a Tecnologia" under the project: FCOMP-01-0124-FEDER-PEst-OE/EEI/UI0760/2011.

References

[1] Friedler, K., Bless, H.: The Information of beliefs at the interface of affective and cognitive processes. In: Frijda, N., et al. (eds.) Emotions and Beliefs: How Feelings Influence Thoughts. Cambridge University Press (2000)

[2] Schwarz, N.: Emotion, cognition, and decision making. Cognition & Emotion 14(4), 433–440 (2000)

[3] Paulus, P., Yang, H.: Idea generation in groups: A basis for creativity in organizations. Org. Behavior and Human Decision Proc. 82, 76–87 (2000)

[4] Paulus, P., Dzindolet, M.: Social influence, creativity and innovation. Social Influence 3(4), 228–247 (2008)

[5] Isen, A., Johnson, M., Mertz, E., Robinson, G.: The influence of positive affect on the unusualness of word associations. Journal of Personality and Social Psychology 48, 1413–1426 (1985)

[6] Hirt, E., Levine, G., McDonald, H., Melton, R., Martin, L.: The role of mood in quantitative and qualitative aspects of performance. Single or multiple mechanisms? Journal of Experimental Social Psychology 33, 602–629 (1997)

[7] Abele-Brehm, A.: Positive and negative mood influences on creativity: Evidence for asymmetrical effects. Polish Psych. Bulletin 23, 203–221 (1992)

[8] Ortony, A.: On making believable emotional agents believable. In: Trapple, R.P. (ed.) Emotions in Humans and Artefacts. MIT Press, Cambridge (2003)

[9] Isen, A., Baron, R.: Positive affect as a factor in organizational behavior. In: Staw, B.M., Cummings, L.L. (eds.) Research in Organizational Behaviour, vol. 13, pp. 1–54. JAI Press, Greenwich (1991)

[10] Frederickson, B.: The role of positive emotions in positive psychology. American Psychologist 56, 18–226 (2001)

[11] Vosburg, S.: Mood and the quantity and quality of ideas. Creativity Research Journal 11, 315–324 (1998)

[12] Laranjeira, J., Marreiros, G., Freitas, C., Santos, R., Carneiro, J., Ramos, C.: A proposed model to include social and emotional context in a Group Idea Generation Support System. In: 3rd IEEE International Conference on Social Computing, Boston, USA (2011)

Thermal Comfort Support Application for Smart Home Control

Félix Iglesias Vázquez and Wolfgang Kastner

Abstract. The main goal of *smart home* applications can be summarized as improving (or keeping) users' comfort maximizing energy savings. Holistic management and awareness about users' habits play fundamental roles in this achievement, as they add predictive, adaptive and conflict resolution capabilities to the home automation system. The present paper presents an application for the thermal comfort that utilizes information from habit profiles – occupancy, comfort temperature and comfort relative humidity – and is designed to be integrated within overall smart home approaches. The application pursues to keep outdoor air quality and thermal comfort controlling shading devices and ventilation elements in unoccupied times and exploiting unobtrusively persuasive technologies in occupied periods. This context aware pre-control phase paves the way to minimize heating and cooling costs.

1 Introduction

As far as energy efficiency is concerned, buildings are one of the major target areas adopted by the European Council Action Plan, where reducing cooling and specially heating are considered of prime importance in the coming years [7]. In particular, in the residential sector since, as it is emphasized in the report, space heating takes two thirds of the energy consumption in European residential homes.

In this respect, smart home technologies can provide important improvements on energy reduction and sustainability [10]. Not only optimizing individual elements and appliances, overall approaches where the whole building is considered as an integrated system can entail greater efficiencies [6]. Hierarchical and global management is highly advisable in scenarios where the correct coexistence and even cooperation among multiple applications and services must be guaranteed.

Félix Iglesias Vázquez · Wolfgang Kastner
Vienna University of Technology, Vienna, Austria
e-mail: vazquez@auto.tuwien.ac.at, k@auto.tuwien.ac.at

P. Novais et al. (Eds.): Ambient Intelligence - Software and Applications, AISC 153, pp. 109–118.
springerlink.com

Another outstanding point for the desired optimization is the abstraction of users' behaviors and habits at home. By means of habit patterns, a common underlying basis for home appliances is built, so the global running of the whole habitable space can be optimized (e.g. [4]). In addition, home profiling goes beyond a separated smart home scope, it draws a common field where shared information is utilized to enhance services for neighborhoods or communities [8] as well as to improve building calculations in design phases (e.g. [11] and [21]).

In the present work, a smart home application based on habit profiles for the thermal comfort support is proposed. It pursues to keep the environment as close as possible to the desired air quality and thermal comfort conditions by means of low energy cost strategies taking into account the management of shading devices, automated windows and air dampers. Direct actuation over devices is executed during unoccupied periods, whereas in occupied times actions are substituted by unobtrusive recommendations published in user interfaces or directly on device indicators.

Occupancy profiles are deployed to anticipate future scenarios based on users' habits, whereas comfort profiles (temperature and relative humidity) allow that the system adapts to the users' subjective assessments concerning thermal comfort. The *Thermal Comfort Support Application* is developed under the coverage of global smart home concepts (specifically [24]), and profits from the integration together with other profile-based applications, e.g. [13].

2 Air Quality and Thermal Comfort

Air quality is considered by the European Community policy as one of the major environmental concerns, being the main goal on air pollution the achievement of levels that do not result in unacceptable risks to human health [25]. An acceptable air quality is defined in [2] as:

> (A)ir in which there are no known contaminants at harmful concentrations as determined by cognizant authorities and with which a substantial majority (80% or more) of the people exposed do not express dissatisfaction.

On the other hand, *thermal comfort* is considered as a subjective evaluation from users that depends on metabolic rate, clothing insulation, air temperature, radiant temperature, air speed and humidity [3]. In the ASHRAE55-2004 standard, the subjective evaluation of users concerning thermal comfort is simplified by a PMV model (Predicted Mean Vote) that predicts the mean value of the votes of a large group of persons on the "seven point thermal sensation" scale [23]. As in the ASHRAE standard, many other advanced methodologies deploy PMV models for thermal comfort control [19]. However, in the standard itself it is remarked that: "it may not be possible to achieve an acceptable thermal environment for all occupants of a space due to individual differences, including activity and/or clothing".

Indeed, the Fanger's PMV model has been criticized by some authors as it seems to be unsatisfactory to predict the response of people in the variable conditions of daily life [12, 20]. It is possible to conclude that difficulties appear with the attempt of modeling human opinion under a single model for everybody, regardless of the multiple variables that can cause variations in the thermal comfort appraisal.

Unlike systems based on PMV, *control based on habits* reaches comfort conditions deploying past user adjustments, therefore the system – by means of habit profiles – learns what thermal comfort exactly means for its specific users.[1]

Fig. 1 shows an example of how a possible *air quality and thermal comfort controller based on profiles* would settle the ideal and acceptable zones for air quality and thermal comfort management based on the users' desires (as centers of the ideal zones) and boundaries fixed by competent directives.

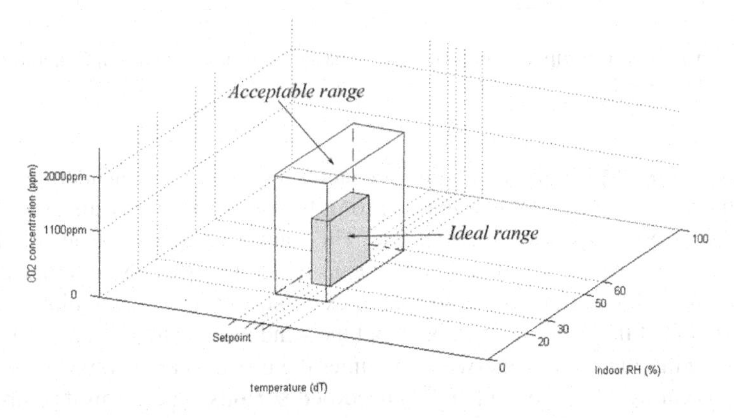

Fig. 1 Example of ideal and acceptable zones for a profile-based system able to manage *comfort temperature habits*. In this case, *comfort relative humidity* ranges are pre-fixed, but could also be dependent on habits. The third parameter, related to air quality, is CO_2 concentration.

We consider temperature and humidity for the thermal comfort, and CO_2 as a pollutant for the air quality. A correct management of these variables guarantees a good final performance. Moreover, note that the *personal factors* emphasized in the standard are partially covered by profiles, as activity and clothing in people usually recur in time and follow daily, weekly and seasonally cadences (habits).

3 Thermal Comfort Support Application

Fig. 2 depicts the schema of the Thermal Comfort Support Application, showing inputs, outputs and included modules/agents. The overall running is explained throughout this section. Regarding the application objectives, it takes two main roles depending on the house occupancy status:

[1] An introduction to the setpoint temperature control based on habit profiles can be found in [13], whereas comparisons of profile-based approaches with other strategies are offered in [24] and [14].

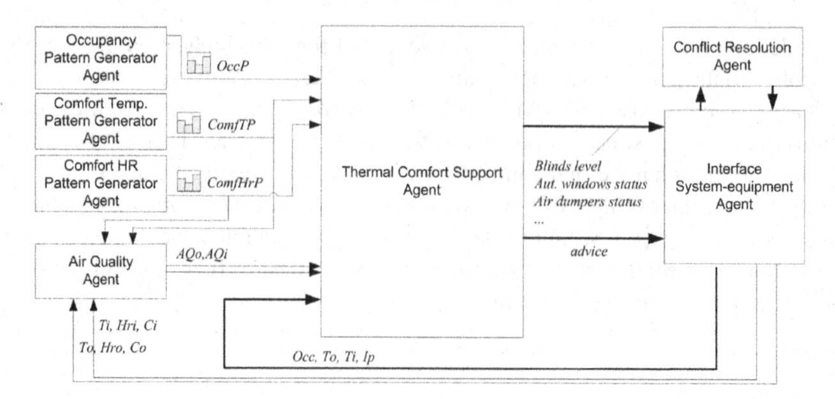

Fig. 2 Thermal comfort support application schema. A description of Pattern Generator modules is published in [13].

In **unoccupied periods**, the controller increases the thermal insulation or balances indoor and outdoor conditions aspiring to maximize indoor air quality and thermal comfort for the next users' arrival. The control strategy is carried out keeping the minimal energy consumption by means of acting on blinds, automated windows, shading devices and air dampers. For example, during summer, blinds can be closed to reflect incident radiation back out the windows and to avoid an additional heating load in the room. Otherwise, in winter the process can be reversed to keep this extra heating. On the other hand, in **occupied periods**, it performs its outputs as advice and recommendations for users, in keeping with the idea of natural comfort with minimum energy costs.

There exist solutions that partially cover some of the intended functionalities; e.g. *smartBLIND*[2] for regulating the incoming light and heat through windows, *NEST* [3] for the thermostat regulation based on habits, or see [16] for proactive home advices. Unlike the presented application, such solutions are usually devised from specific perspectives that have limitations to properly profit the context awareness capabilities of global smart home approaches.

3.1 Occupancy and Times

By means of *occupancy profiles* (*OccP*) and *instantaneous occupancy* (*Occ*), the smart controller not only knows whether there are people at home or not, but also it can predict next presences, absences as well as to know if they will be short or long. Therefore, the combination of *OccP* and *Occ* states four different times: *readiness, presence, short absence and long absence*; in each case, the controller takes a different role, shown in Table 1.

[2] http://www.smartblind.com/
[3] http://www.nest.com/

Table 1 Actuation of thermal comfort support application according to the type of time.

Time	Role	Comments
Readiness	Active	Management of devices (blinds, dampers, shading, etc.)
Presence	Advisor	Outputs in form of messages for users
Short absence	Active	Management of devices (blinds, dampers, shading, etc.)
Long absence	Off	No action

Regardless of the fact that cooling and heating is used in short unoccupied times or not,[4] the Thermal Comfort Support Application tries to approach to desired thermal conditions by means of low energy demanding block/unblock strategies. The energy cost differences between performances with and without shading devices (internal and external) have been measured for the Canadian housing in [9]. They conclude with about 15-20% of energy savings after an appropriate management of shading devices, both in winter and in summer seasons.

3.2 Air Quality Assessments

Air quality modules perform evaluations of the current indoor and outdoor air quality. For the assessments, some limits are settled from the comfort values given by the profiles until rates laid down in standards in order to guarantee habitable conditions.

A suitable way to obtain the qualitative output is by means of fuzzy inference systems.[5] The difference of comfort and real values in temperature and relative humidity ($ComfTP$, $ComfHrP$; T_o, T_i; Hr_o, Hr_i), as well as CO_2 concentration (C_o, C_i) are normal inputs, whereas air quality is the output (AQ_o, AQ_i).

$$dTemp(t) = ComfTP[n] - T(t) \tag{1}$$

$$dHr(t) = ComfHrP[n] - Hr(t) \tag{2}$$

In equations 1 and 2, n and t denote equally time. Whereas t stands for instantaneous measures, n marks discrete time periods fixed by the profiles time-resolution. Irrespectively of n and t, the polling period used by the system to execute actions is a parameter that depends on the final scenario and the building characteristics (on the other hand, a change in the type of time is considered as a *interruption*).

Instead of directly managing temperature and humidity values, air quality modules compare them with comfort values stored in profiles. Therefore, the system, according to the user habit for every moment of the day, adjusts dynamically the ideal point for the later fuzzy reasoning.

[4] Depending on the building/house location, changing to setback or setup heating/cooling modes during absences is a common practice recommended by experts, e.g. [22].

[5] The suitability of fuzzy systems for air quality can be seen, for example, in [5].

3.3 Decision Taking

Based on the indoor and outdoor air quality assessments and considering that we are in readiness or short absence times, a simplified version of the decision taking process can be appreciated in Table 2.

Table 2 Air quality insulation table of rules (summarized version from a fuzzy inference system).

Indoor AQ	Outdoor AQ	Action
good	good	idle
good	tolerable	insulate
good	bad	insulate
tolerable	good	balance
tolerable	tolerable	idle
tolerable	bad	insulate
bad	good	balance
bad	tolerable	balance
bad	bad	idle

The meaning of the actions is as follows: when *idle*, the smart control does not execute any action; with *balance* it opens blinds, automated windows, air supply and exhaust dampers; whereas *insulate* means to close blinds, automated windows, air supply and exhaust dampers.

However, there can happen some controversial situations. For instance, it is possible that ventilation insulation is required to keep good indoor conditions but solar gain through panes can even improve the indoor air quality. In order to solve such conflicts, *solar gain through windows* (Q_{sg}) and *sensible transmission through glass* (Q_{tg}) are compared. Q_{sg} and Q_{tg} can be summarized [1] in the next equations:

$$Q_{sg} = k_1 \cdot A_g \cdot SHGC \cdot SC \cdot I_p \tag{3}$$

$$Q_{tg} = k_2 \cdot A_g \cdot U_g(T_o - T_i) \tag{4}$$

where A_g stands for the pane's surface, U_g is the U-Factor or U-Value and indicates how well the window conducts heat. On the other hand, $SHGC$ is the Solar Heat Gain Coefficient (dimensionless) and measures how well a product blocks heat from the sun. SC is the Shading Factor (dimensionless) and I_p offers the solar radiation over the window [W/m^2]. k_1 and k_2 represent engineering parameters.

The comparison between equations 3 and 4 give an appraisal about the thermal gains through the windows in their normal status (blinds[6] rolled), and help the system to know if acting through the blinds can be suitable or not. For this purpose, Table 3 is defined for the operation of the blinds according to the thermal gain contribution (it has priority over Table 2, only conflictive cases are shown).

[6] From now on, we refer indistinctly all kind of shading devices as blinds.

Table 3 Set Device Status Process Table for Blind Operation.

Desired effect	Q_{tg} sign	$\lvert Q_{tg}\rvert ... \lvert Q_{sg}\rvert$	Blinds/shutters
cool down	- (loss)	$>$	open
cool down	-	$<$	close
warm up	-	$>$	close
warm up	-	$<$	open

The *desired effect* is calculated comparing the indoor/room temperature (T_i) with the setpoint temperature offered by profiles for the specific moment ($ComfTP[n]$). For the case $T_i = 12°C$ and $ComfTP[n_0] = 20°C$, the system realizes that *warming up* strategies are suitable at the moment. Q_{tg} sign marks if there are thermal conduct gains or losses through the window; since Q_{sg} cannot be negative, conflicts can happen when Q_{tg} is negative.

The simple approach given by equations 3 and 4 can be enough to have suitable assessments about heat gains and losses due to conduction-convection and solar radiation at the same time. In this point, it is necessary to count on additional information about the house or building. For example, accurate SC and $SHGC$ assessments take different values depending on materials, type of fenestration, type of room floor, date and time, building location and geometry, glazing orientation, solar incidence angle, frame effects, shading devices, etc. By means of published tables [1] and a set of sensors, smart systems can perform adequate calculations.

Nevertheless, beyond the proposed solution, the usage of simulation tools to predict different future scenarios and take decisions ahead (building simulations for decision taking of controllers) is an advanced option to consider; indeed, the agreement between measures and simulations for solar gain in windows have been already demonstrated [18]. In addition, due to the exposed dependence on the building characteristics, smart home control supported by Building Information Models (BIM) [17] is empowered by this sort of applications.

3.4 Advice Generation

The *advisor capabilities* of the smart system are intended to produce advice and recommendations to users in occupied times. The basic idea follows the principles exposed by Intille in [16], where the house of the future explores the persuasive capabilities of technology by presenting information at precisely the right time and place. The final design avoids unsatisfactory conflicts between users and devices – instead it helps inhabitants to learn how to control the environment on their own.

Therefore, the actuation of the Thermal Comfort Support Application is substituted by the publication of its *reasonings* in the form of advice. It covers mainly: (a) Opening and closing windows and hatches. (b) Applying and removing internal and external shading devices, but considering the users' lighting requirements.[7]

[7] This demands the integration of indoor and outdoor light level sensors. Lighting level profiles can also be utilized for the final advice performance.

Advices are published in the home via user interfaces or also directly in the objective house elements; for example, dedicated color LED lights on windows and blinds can inform about the convenience of changing the status. This informative service can perfectly coexist and even empower other persuasive applications for the comfort and energy optimization, e.g. [15].

4 Application Example

The Thermal Comfort Support Application can be clearly assessed in the next example cases. The two cases can be considered consecutive in time to understand how the system adapts to a sudden scenario change.

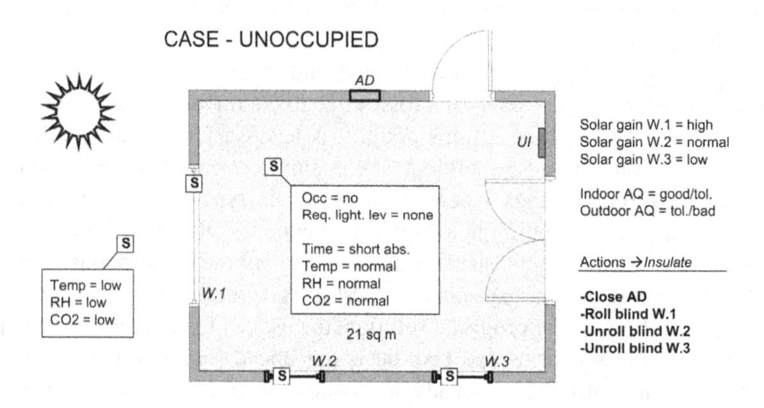

Fig. 3 Example of the *thermal comfort support* application: unoccupied case.

In Fig. 3, by means of *Occ* and *OccP* the system knows that the present *type of time* is *short absence*. The weather corresponds to a winter season, with outdoor low temperatures but with a sunny day. The fuzzy evaluations concerning air quality lead the system to take the *insulation* strategy; it is, to close the air damper (AD) and the blinds. However, solar gains through *W*.1 overcome thermal losses so the blinds in this window keep unrolled.

In Fig. 4, some users enter the room and the application leaves the active role to switch to advisor mode. Air dampers are opened by the normal HVAC controllers. Depending on the customized configuration, the status of blinds can remain the same (according to the decisions taken in the absent period) or can be restored to the conditions saved during the last occupancy. In any case, if the status does not coincide with the ideal scenario reasoned by the system, messages may be displayed at user interfaces or useful information may be indicated directly at particular devices. Note that, in the example, there is no advice concerning the window W.2 due to certain level of admitted uncertainty as lighting requirements also come into play.

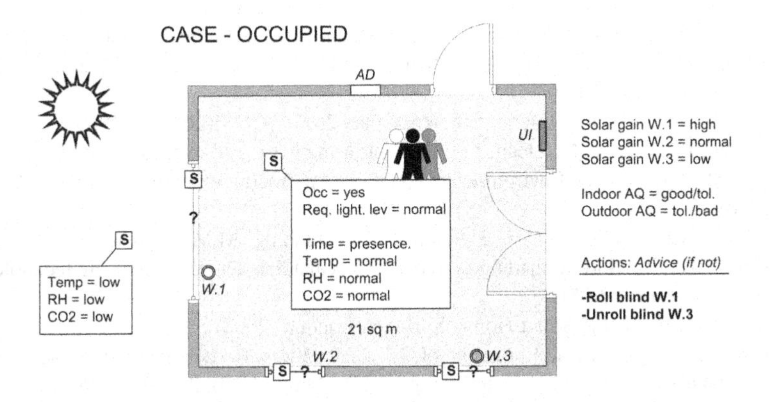

Fig. 4 Example of the *thermal comfort support* application: occupied case.

5 Conclusions

The combination of holistic smart home control and users' habit awareness offers valuable applications for the energy consumption minimization. Home services must be understood as coexisting in a common environment that evolves depending on the users' requirements. Superpositions, conflicts and transitions among applications happen continuously, users' habits draw connections and patterns among them and overall control can manage them to optimize the global running.

The Thermal Comfort Support Application fits this idea; it is independent of other controllers, devices and machinery, although it obviously requires a fluid communication and hierarchical arrangement of control modules in order to correctly manage possible conflicts, shared resources and information. In addition, transverse services that consider users' behaviors are necessary to make flexible and smoothly glue the often low fluid relationship between users and home automation systems.

Acknowledgements. The presented work was funded by the HdZ+ fund of the Austrian Research Promotion Agency FFG under the project ID 822170.

References

1. ASHRAE. Chapters 27 to 29, ASHRAE Handbook – Fundamentals (1997)
2. ASHRAE. Ventilation for Acceptable Indoor Air Quality. Standard 62-2001 (2001)
3. ASHRAE. Thermal Environmental Conditions for Human Occupancy. Standard 55-2004 (2004)
4. Barbato, A., Borsani, L., Capone, A., Melzi, S.: Home energy saving through a user profiling system based on wireless sensors. In: Proceedings of the First ACM Workshop on Embedded Sensing Systems for Energy-Efficiency in Buildings, Berkeley, California, pp. 49–54 (2009)
5. Bernard, Fisher: Fuzzy environmental decision-making: applications to air pollution. Atmospheric Environment 37(14), 1865–1877 (2003)

6. Borggaard, J., Burns, J.A., Surana, A., Zietsman, L.: Control, estimation and optimization of energy efficient buildings. In: American Control Conference, ACC 2009, pp. 837–841 (June 2009)
7. European Commission. Energy Efficiency Plan 2011 - COM(2011), 109 final
8. Crosbie, T., Dawood, N., Dean, J.: Energy profiling in the life-cycle assessment of buildings. Management of Environmental Quality: An International Journal 21(1), 20–31 (2010)
9. Galasiu, A.D., Reinhart, C.F., Swinton, M.C., Manning, M.M.: Assessment of energy performance of window shading systems at the canadian centre for housing technology. Technical report (2005)
10. Harper, R.: Inside the Smart Home. Springer, London (2003)
11. Hoes, P., Hensen, J.L.M., Loomans, M.G.L.C., de Vries, B., Bourgeois, D.: User behavior in whole building simulation. Energy and Buildings 41(3), 295–302 (2009)
12. Humphreys, M.A., Nicol, J.F.: The validity of iso-pmv for predicting comfort votes in every-day thermal environments. Energy and Buildings 34(6), 667–684 (2002)
13. Iglesias, F., Kastner, W.: Usage profiles for sustainable buildings. In: 2010 IEEE Conference on Emerging Technologies and Factory Automation (ETFA), pp. 1–8 (2010)
14. Iglesias, F., Kastner, W.: Clustering methods for occupancy prediction in smart home control. In: 2011 IEEE International Symposium on Industrial Electronics (ISIE), pp. 1321–1328 (June 2011)
15. Iglesias, F., Kastner, W., Cantos, S., Reinisch, C.: Electricity load management in smart home control. In: 12th International IBPSA Conference, Building Simulation 2011, pp. 957–964 (2011)
16. Intille, S.S.: Designing a home of the future. IEEE Pervasive Computing 1(2), 76–82 (2002)
17. Kofler, M.J., Kastner, W.: A knowledge base for energy-efficient smart homes. In: 2010 IEEE International Energy Conference and Exhibition (EnergyCon), pp. 85–90 (December 2010)
18. Kotey, N.A., Wright, J.L., Collins, M.R., Barnaby, C.S.: Solar gain through windows with shading devices: Simulation versus measurement. ASHRAE Transactions 115(2) (2009)
19. Liang, J., Du, R.: Thermal comfort control based on neural network for hvac application. In: Proceedings of 2005 IEEE Conference on Control Applications, CCA 2005, pp. 819–824 (August 2005)
20. Santamouris, M.: Solar thermal technologies for buildings: the state of the art. James & James (Science Publishers), London (2003)
21. Mahdavi, A., Pröglhöf, C.: User behaviour and energy performance in buildings. In: IEWT 2009, Int. Energy Economics Workshop TUV, pp. 1–13 (2009)
22. Moon, J.W., Han, S.-H.: Thermostat strategies impact on energy consumption in residential buildings. Energy and Buildings 43(2-3), 338–346 (2011)
23. Fanger, P.O.: Thermal comfort. Danish Technical Press, Copenhagen (1970)
24. Reinisch, C., Kofler, M.J., Iglesias, F., Kastner, W.: Thinkhome: Energy efficiency in future smart homes. EURASIP Journal on Embedded Systems 2011, 18 pages (2011)
25. Scientific Committee on Health and Environmental Risks, SCHER. Opinion on risk assessment on indoor air quality. European Commission, Health & Consumer Protection Directorate-General, Brussels (2007)

Management Platform to Support Intellectually Disabled People Daily Tasks Using Android Smartphones

Fernando Jorge Hernández, Amaia Méndez Zorrilla, and Begoña García Zapirain

Abstract. People with intellectual disabilities live in supervised flats in an autonomous way and need daily care hours for carrying out certain specific tasks.

This paper presents a technological solution to promote independent living for people with intellectual disabilities. The aim is to develop an aid that reminds the user about the most important tasks they have to perform.

The device chosen to convey this information is an Android Smartphone. The social workers, psychologists and users involved in its development have assessed the application very positively, as it meets the needs of this disabled collective perfectly.

Keywords: Smartphone, Tasks Reminders, Intellectually disabled people, Android.

1 Introduction

The disabled face numerous difficulties in their daily lives when it comes to living independently and carrying out their household chores, travelling to their workplace, taking medication, and the like. In a study by the National Institute of Statistics in the year 2008, it was revealed that 873,000 people aged between 16 and 64 -and who lived at home- possessed some kind of disability certificate [1].

The most common disability types found among the population in 2008 were those associated with osteoarticular (18.5%) and intellectual (17.6%) deficiencies [2]. And these people are precisely the beneficiaries of this project focused on situations in which intellectually disabled people can lead an independent life.

Fernando Jorge Hernández · Amaia Méndez Zorrilla · Begoña García Zapirain
DeustoTech-Life Unit, Deusto Institute of Technology, University of Deusto,
Avda. de las Universidades, 24. 48007 Bilbao, Spain
e-mail: {fernandojorge,amaia.mendez,mbgarciazapi}@deusto.es

P. Novais et al. (Eds.): Ambient Intelligence - Software and Applications, AISC 153, pp. 119–128.
springerlink.com

There are many programs (such as Etxeratu, promoted by the Down Syndrome Foundation of the Basque Country [3]) in which various intellectually disabled people from collectives or foundations live independently in supported accommodation, either sharing with other people having the same disability or alone. These people perform their household chores, like doing the washing, ironing or cooking by themselves, as well as other day-to-day tasks related to independent living, like paying their electricity and water bills.

Currently in the supervised flats, people with disabilities who reside in them, have a few hours a week support from coordinators of their own organizations. They also have other methods to remind them tasks to perform, such as, shared whiteboards.

These problems are intended to be reduced with the use of the proposed platform. The tasks can be included on the website individually, set specific times, order, and set how often to be repeated (ie, daily, weekly, working days, etc.).

And, the mobile device is a more effective way of communicating with people with disabilities without having to watch for the state of tasks currently in use.

Thanks to this project, a system has been produced that reminds them to carry out their corresponding tasks (such as leaving for work, taking a particular medicine or putting the washing machine on) in a simple way through the use of a mobile device.

Among the group of people with intellectual disabilities the use of these devices is widely spread, such that 89.7% of them have mobile phones and the 59.3% uses it regularly. [4]

2 Objectives

The main aim of this project is the development of a platform that enables those responsible for organizations to send various warnings or alerts to disabled users that live independently in supported accommodation. In order to attain this goal, the authors divided these aims into three categories: technical, social and evaluation.

2.1 Technical Objectives

The main technical aims of this project are to:

- Design and develop a platform capable of managing an individual calendar for each disabled person assigned to the system in such a way that the alerts are inserted with ease and efficiency. An alert is normally composed of a text accompanied by an image (Fig. 3).
- Design permits and roles that lead to the independence of the intellectually disabled and distinguish one organization's monitors form another's, so that the platform can be used simultaneously by different organizations.
- Design and develop a mobile application capable of synchronization with the platform and displaying the alerts in a straightforward way for the user

carrying the device, without the user having to know how to operate the mobile device.

- Ascertain whether the user receives and accepts the alerts sent and check that the tasks have been carried out by obtaining the result/number of alert tasks performed in comparison with the total number of alerts.

2.2 *Social Objectives*

The intended social objectives of this project are to:

- Foster disabled people's independence and safety, providing them with help in their day to day activities.
- Improve e-inclusion of the disabled, that is, bring new communication technologies (as a means of support) closer to this collective.
- Boost the confidence of disabled people's parents or guardians so that they can include the most important tasks in the calendar.

2.3 *Evaluation Objectives*

Gathering results are:

- Whether the user receives and accepts the alerts.
- The project´s advantages.
- Compare the system with others system as Google Calendar.
- Analysis of the most common types of alerts.

3 Proposed System Design

The platform design was divided into two quite different blocks (Fig. 1). On the one hand, there is the part corresponding to the device employed, which is in charge of maintaining synchrony with the web platform and collecting the alerts that have been specifically added to this device.

On the other hand, we have the web platform's server, which facilitates the display of all the system's user calendars and the addition of new alerts to each of the former. Communication between the mobile device and the web platform is performed through the communication module, thus concealing the details of sending and receiving data from the rest of the system.

3.1 *Smartphone Application*

The application works in the following way:

Once the device has been switched on, it is safely connected to the web platform; once authorized, it downloads the alerts saved -these alerts are specifically composed for an exact time and a URL (Uniform Resource Locator) to an image stored in the server- and ends the connection. Thus, the application continues running even if the device has no Internet connection, enabling total system availability. [5]

Every time the control module detects a saved alert for a particular moment, the text and image relating to the alert are displayed on the device and it waits for the response from the user accepting the alert.

When the user accepts, a notification is sent to the server informing that the user has accepted that particular alert. If five minutes go by without the user accepting the alert, the server registers that it has not been accepted or received correctly.

The device application was designed in modular form, there being four specific modules:

- Database Module: The alerts to be generated are stored in the device through this module. Only pending alerts are stored as the previous ones are registered in the server.
- Interface Module: This module is in charge of generating the graphic interface in the device. Therefore, providing solely the text and image, the rest of the application is hidden from development of the interface.
- Calendar Control Module: This module deals with the application's logic. It reviews the alerts stored in the device and generates the events necessary for the interface creation. In addition, it also receives the user's interaction with the interface and returns the response to the communication server.
- Communication Module: This module establishes communication with the server in the three main tasks: authentication, alert receipt and forwarding confirmation of a specific alert).

3.2 Server Application

The functioning/operation of the server is divided into two parts. On the one hand, there is the web platform, which provides the possibility of creating various users, associating devices with the system and managing the calendars; this enables the platform user to handle alerts in a straightforward way. On the other hand, the platform communicates with the associated devices by sending them the new alerts added to the system and by receiving confirmation that the alerts have been accepted by the device users.

The modules making up the server are the following:

- Communication Module: This allows communication with the device, hiding this functionality from the rest of the system.
- Roles Module: This is one of the system's most important modules. Because of the information it contains, such as the users' personal details, daily routines, etc., it is vitally important that this module cannot be accessed by other users.

  For this reason the four following profiles with different characteristics were created System Administrator, Coordinator, Medium and Low User.

A diagram (Fig. 2) of these profiles can be seen below:

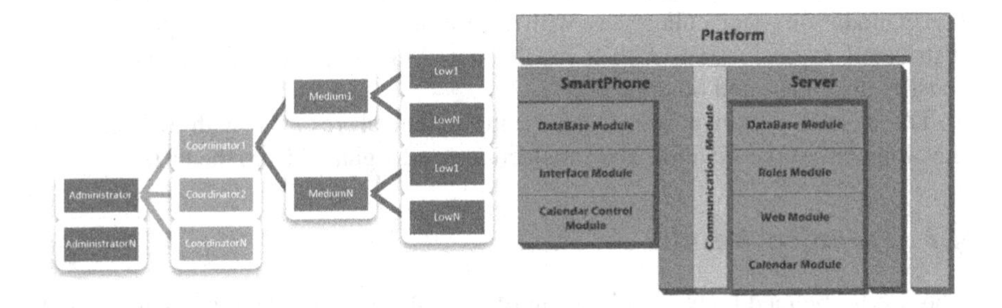

Fig. 1 High level Diagram of Platform **Fig. 2** Profile Diagram

- Database Module: This stores all the alerts added to the application, including whether they have been notified of and whether accepted by the user carrying the device.
- Web Module: By means of Struts 2, a web platform enabling the user to interact simply with the system was developed. The following operations, among others, can be carried out through the web user management, device management and calendar of users management.
- Calendar Module: This module runs a calendar for each user, adapts that calendar and enables the insertion of new alerts, which it also edits [6]. When it comes to creating a new alert, the following need to be included an alert text, an image related to the task and it can be programmed to recur daily, weekly, monthly, etc.

4 Results

Currently, the system is in an advanced stage of development. They have been implemented the three main parts, server alerts, Web interface and mobile application for Android OS devices, but for future tests will use other developments Windows Phone 7 [7] IOS [8] or BlackBerry [9].

Using a mobile device smartphone is due on the one hand the great potential of development with these devices today, and otherwise completely normal in their use among any sector of the population [10]. Thus, it prevents the user from having to carry another dedicated device just for this function, devices such as bracelets, personal trackers...

To develop the system has been used an agile methodology, this is Extreme Programming [11] [12].

This approach allows each iteration performed with new features or changes, carried out tests allowing users to see their reaction to possible changes or bug fixes.

To check the operation of the platform, potential changes, and acceptance by users of the system has been a previous pilot.

Collaboration on this project was received from an organization that runs a specific program for the independent living of disabled people residing in rented accommodation. The collectives with disabilities involved in this program are distributed as such: 68% with a intellectual disability, 26% with a physical or sensory disability, and others 6%.

The results obtained are divided into technical results –with reference to development of the system-, and satisfaction results gained from the tests.

4.1 Technical Results

The result obtained is the development of a platform divided into two parts: a web part that deals with the management of users, devices and calendars, and the other part is with the device application (Fig.3).

As regards the client application, the system was developed for devices with the Android operating system [13]. The reasons for this choice are its widespread diffusion, the great number of devices that use it (both brand marks and types/models) and its easy programming [14] in comparison with other similar systems like Windows Phone 7 or IOS (*Iphone Operative System*).

Use of the mobile device was simplified by the application's ability to start up automatically when the device is switched on. Therefore, the user does not have to check whether the application is working properly.

Once the trials with real users had been carried out, the technical results obtained were satisfactory, though it should be pointed out that certain factors during the trials provided food for thought, such as:

- The user switches on the device but does not introduce the PIN, and so the application does not start up.
- The device's internet connection. Should a service error or any other similar error occur in the platform, the system sends an alert to the coordinators because it has detected that a disabled person has not accepted any alerts for a whole day.

4.2 Tests Results

Once the testing with real users had been completed, a study was undertaken to analyze the most common type of alerts in the platform. The results (Fig. 4) indicate that most alerts are related to the work schedule, that is, at what time they should leave for work. Many others refer to such domestic chores as cooking, cleaning, doing the washing, ironing, etc. To a lesser extent, other alerts concerning medication, appointments with the doctor or specialist are also common.

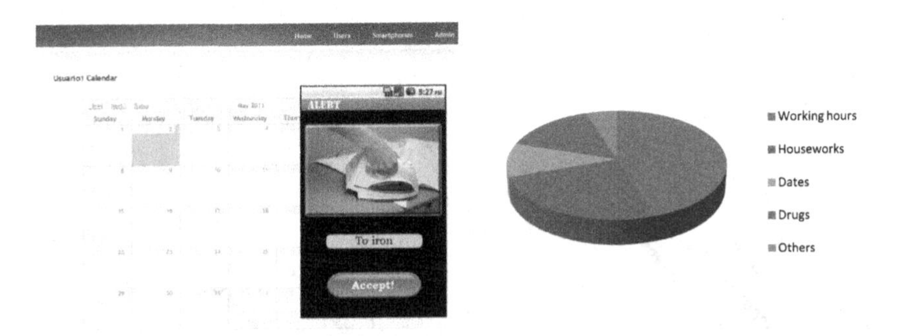

Fig. 3 Web and Smartphone App Example **Fig. 4** Results Diagram

Besides these results, other factors such as users alerts acceptance, the real advantages after having use the system or the devices that best adapt to the users' characteristics have been analyzed.

With regard to users' acceptance rate, has been deemed as very positive. 90% of people with intellectual disabilities involved in piloting have received all their corresponding alerts daily, accepting and performing them immediately (Fig. 5).

After having analyze the alert's acceptance rate in the application, the information has been contrasted with the people responsible for each disabled person, so significant improvements have been identified in some aspects such as punctuality going to work or to their center, a better control of medicines taken or in making houseworks.

The remaining 10% have perceived as unnecessary the alert for houseworks or appointments, although the use of the tool to let them know when to leave work or when to take medicines have been fully accepted.

After the trials undertaken with real users, a comparative study between this platform and Google Calendar [15] was carried out, due to the fact that this is currently the most popular calendar service. It also has its own application in Android mobile devices. The most important reasons why a new platform needed to be developed are:

- Problems with privacy of data.
- Complexity for people with intellectual disabilities when consulting the alerts on their mobile devices (due to the fact that it is necessary to know how the calendar and its different functions work beforehand).
- The new platform produced makes it possible for only the relevant information to be displayed, concealing other complexities from the user's eyes.

Another key aspect when measuring the success of the pilot has been user satisfaction with the device, both in handling and in use, when it is used by people with disabilities.This has been done several surveys with the following results (Fig. 6)

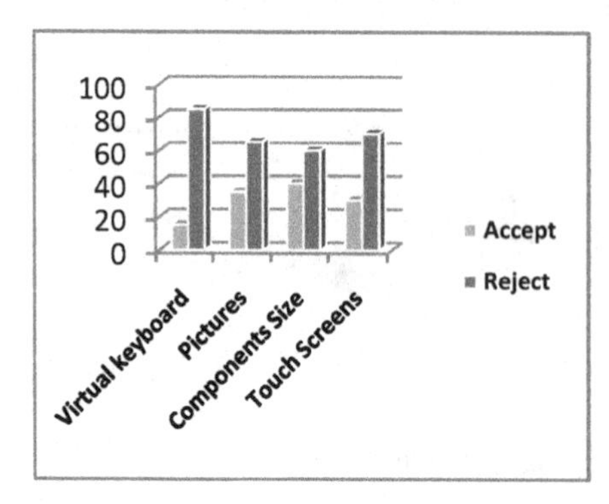

Fig. 5 Results total alerts Diagram

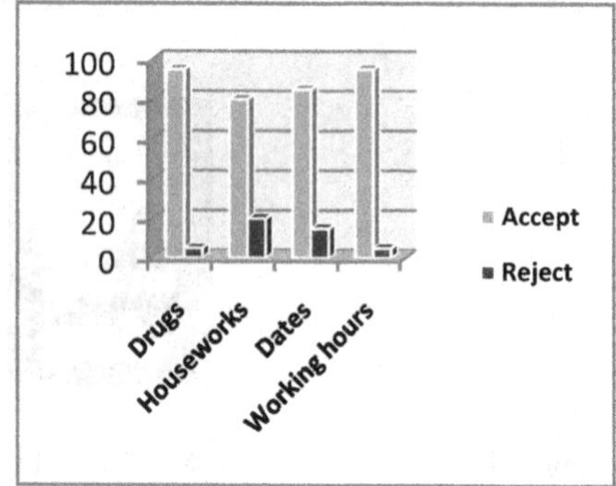

Fig. 6 Users`Survey Results

Thus, we can draw some conclusions mean significant changes to the choice of the device for each user, and most importantly, the redesign of the interface of the application. [16] [17]

In these tests we have used various devices to test user interaction with the different characteristics of each device. (Table 1. Mobile Characteristics)

Table 1 Mobile Characteristics

Mobile Phone Type	Screen Size	Touch Screen	Physical Keyboard
1	7”	●	
2	3.7”	●	●
3	10.1”	●	
4	4”	●	
5	3.2”		●

70% of users prefer a keyboard device with a real rather than virtual.

Against technological progress, existing touch screens are a handicap for people with disabilities when using the device [18] [19].

In the application interface has been detected also two important aspects:

- Size of the components of the interface. Buttons, text boxes and images should be presented with a much larger size than usual, and always following a simple order.
- The images presented in these applications must be real images, ie images such as laundry, mopping, vacuuming, etc., are snapshots of reality. The use of drawings, sketches or diagrams can lead to confusion. Although, for this is necessary to configure the system previously.

5 Conclusions

The outcome of the project is very satisfactory as it allows a group of disabled people, among other things to, carry out their daily lives, do the housework, and travel from their homes to the workplace punctually, completely autonomously. Although it´s necessary, coordinators to configure the system previously.

Moreover, the use of new technologies has been adapted to the disabled, to be a technology well accepted by this group, paving the way for new projects in the sector, exploiting the characteristics of these devices such as GPS, accelerometer...

In the future, the use of this platform may be expanded to other collectives, such as the elderly or sufferers of Alzheimer's, may even be perfectly useful for people without any disability, and this platform may be also introduced to other devices with Windows Phone, iOS and BlackBerry.

Acknowledgment. This work was partially supported by the Basque Country Department of Education, Universities and Research, as well as the Basque Country Department of Innovation and Industry. We would also like to express our gratitude to the Down Syndrome Foundation of the Basque Country, GAIA, Dinitel, Discusland and Thaumat, for their help and support.

References

[1] Instituto Nacional de Estadística: El empleo de las personas con discapacidad (2010)
[2] Margallo, F., Gil, C., Rubio, E., Jiménez, E.: Valoración, Orientación e Inserción Laboral de Personas con Discapacidad. Método Estrella (2006)
[3] Fundación Síndrome de Down del País Vasco (WWW Document), http://www.downpv.org/
[4] Gutierrez, P., Martorell, A.: People with Intellectual Disability and ICTs (2011), http://www.revistacomunicar.com
[5] Kemp, R., Palmer, N., Kieldmann, T., Bal, H.: Cuckoo: A Computation Offloading Framework for Smartphones (2010)
[6] JMonthCalendar (WWW Document), http://www.bytecyclist.com/projects/jmonthcalendar/
[7] Lee, H., Chuvyrov, E.: Windows Phone 7 development. Apress (2010)
[8] Zdziarski, J.: iPhone SDK Application Development. O'Reilly (2009)
[9] Salter, G.: Convergence of synchronous and asynchoronous. Learning Technologies (2005)
[10] 1.1 Billion Smartphones By 2013 (WWW document), http://www.i4u.com/29160/11-billion-smartphones-2013
[11] Abrahamsson, P., Marchesi, M., Maurer, F.: Agile Processes in Software Engineering and Extreme Programming. In: 10th International Conference (2009)
[12] Maruping, L., Venkatesh, V., Agarwal, R.: A control theory perspective on agile methodology use and changing user requirements. Information Systems Research (2009)
[13] Hashimi, S., Komatineni, S., MacLean, D., MacLean, D.: Pro Android 2, 1st edn. (2010)
[14] Michael, H.: Pro Android Best Practices. 1st edn. (2011)

[15] Google Calendar Privacy Policy (WWW Document), `http://www.google.com/intl/es/privacy/privacy-policy.html`

[16] Balagtas-Fernandez, F., Hussmann, H.: Evaluation of User-Interfaces for Mobile Application Development Environments. In: Jacko, J.A. (ed.) HCII 2009, Part I. LNCS, vol. 5610, pp. 204–213. Springer, Heidelberg (2009)

[17] Balagtas-Fernandez, F., Hussman, H., Forrai, J.: Evaluation of User Interface Design and Input Methods for Applications on Mobile Touch Screen Devices (2009)

[18] Arroba, P., Vallejo, J.C., Araujo, Á., Fraga, D., Moya, J.M.: A Methodology for Developing Accessible Mobile Platforms over Leading Devices for Visually Impaired People. In: Bravo, J., Hervás, R., Villarreal, V. (eds.) IWAAL 2011. LNCS, vol. 6693, pp. 209–215. Springer, Heidelberg (2011)

[19] Konig, A., Radle, R., Reiterer, H.: Interactive design of multimodal user interfaces, Reducing technical and visual complexity. Journal Multimodal User Interfaces (2010)

Family Carebook: A Case Study on Designing Peace of Mind for Family Caregivers

Matty Cruijsberg, Martijn H. Vastenburg, and Pieter Jan Stappers

Abstract. Family caregivers often are key to aging-in-place. They not only help with everyday activities and needs, but also tend to be a valuable part of the senior's social network. Many family carers however experience a hard time combining care activities with their busy job and family life. At the same time, they indicate they feel insufficiently aware of the actual situation and needs of the senior, which results in worries and inefficiencies in care. Present communication systems appear to be inadequate in addressing these information needs of family carers. Towards improving awareness of family carers, and ultimately enabling peace of mind, new design directions need to be explored. This paper presents a design exploration in which an interactive communication concept was developed. The focus was on supporting the family carer. A working prototype was studied in the field. The findings from the case study suggest that both functional awareness and peace of mind need to be considered towards creating new communication services for a sustainable family care setting.

Keywords: Design exploration, peace of mind, aging-in-place, family care.

1 Introduction

The number of senior citizens in western societies is expected to increase dramatically in the next decades. Seniors often prefer to live independently in a familiar setting as long as possible. Family carers play a central role in supporting seniors when they need help living on their own. Often family carers combine family care with a paid job (71% in The Netherlands [SER, 2007]), which results in busy schedules. To provide the right care at the right time, caregivers need to be aware

Matty Cruijsberg · Martijn H. Vastenburg · Pieter Jan Stappers
Faculty of Industrial Design Engineering, Delft University of Technology,
Landbergstraat 15, 2628 CE, Delft, The Netherlands
e-mail: mcruijsberg@gmail.com,
{m.h.vastenburg,p.j.stappers}@tudelft.nl

P. Novais et al. (Eds.): Ambient Intelligence - Software and Applications, AISC 153, pp. 129–136.
springerlink.com

of the actual needs of seniors. Interviews conducted in this study do however indicate that the current level of awareness is insufficient.

ICT systems that provide awareness of the actual situation of the senior could potentially support family carers in providing care at the right moment and provide peace of mind. A wide range of systems with the aim of assisting seniors and their carers has already been developed. Sensors are increasingly being used in the homes of seniors to detect for example presence, the key application appears to be monitoring and alarming. Albeit valuable, present systems do however not properly address the needs of carers in terms of providing peace of mind.

To better understand how awareness systems can be used to enable peace of mind for family carers, a design exploration was conducted. The specific needs of the family carer are mapped and design directions are explored. The present paper describes the design steps that resulted in a prototype awareness display for family carers and the findings from a field trial. The case study serves as an example of how awareness systems could be designed to improve both functional awareness and peace of mind.

2 Related Work

The use of ICT to support family carers has been studied before. A key example is Digital Family Portrait (DFP), which was designed to provide awareness and peace of mind to extended family members through showing trends in a person's daily life [7]. DFP is based on the assumption that peace of mind can be provided by monitoring day-to-day activities. Icons show information about information categories over time. DFP visualizes plain aggregated sensor data. A field trial showed that the DFP ambient display was appreciated as a way to learn about the life of a distant family member. The information shown on the prototype was based on user reports rather than real sensors. One might argue that the information displayed by DFP cannot be collected using sensors alone. Research is needed to find ways to collect the required information.

Similar to DFP, CareNet Display [2] presents information on the elder's condition. Users can zoom out from detailed information on events to a 5-day trend view. Whereas DFP could only be used as a source of information, CareNet Display enables carers to coordinate care tasks. A field trial showed that local carers experienced lower stress levels [2]. The exit-interviews did however indicate that the system lacked a 'human touch': the sensor data did provide a quantitative view, whereas carers felt the need to know qualitative view.

The ASTRA project [5] focused on *sharing* to increase family connectedness. Human-human communication was used to detect the needs and state of a caretaker. A mobile device was used to share information on the go. Whereas information sharing seems to be a proper way to enable social communication, it is unclear if people are willing and able to share information regarding care needs in a similar way. In exploring design directions, sharing could be considered as a mechanism to exchange information regarding care.

The challenge of combining user input with sensor data has been studied in the Daily Activities Diarist (DAD) project [6]. The DAD application was used to journal activities of daily life. Seniors were asked to correct the activities detected by the system, eventually enabling the system to automatically recognize daily activities. The journal was found to have a positive effect on social connectedness, and was used as a peripheral social awareness cue [6]. Since DAD provides only a description of activities rather than a subjective view, one might expect DAD not only to reassure carers, but also to raise new questions.

The ContextContacts [4] and FamilyAware [8] projects focused on sharing context information using mobile phones. Both systems could potentially be used to support a group of carers to coordinate care tasks and determine availability. Whereas ContextContacts was limited to sharing location information, FamilyAware did include the option to communicate. Even though these systems could be used to increase awareness of presence and availability, none of the systems aimed to increase awareness of actual care needs.

To conclude, the use of communication devices and sensor systems to support family care has been explored extensively. Several projects focus on communication, resulting in improved social connections, while other projects emphasize the use of sensors to automatically detect the present state or patterns in time. Based on explorations with family carers, we do however expect that combination of monitoring *and* communication will outperform existing systems in terms of peace of mind and awareness. The present paper presents the findings from a case study in which the needs of family carers were studied, and a hybrid communication and monitoring system was developed. A working prototype was deployed in the field, in order to collect data on user experiences in a realistic setting. The paper concludes with the findings, discussion and suggestions for future work.

3 User Research

To better understand the current family care situation, and to understand how family carers could be supported using ICT systems, a contextmapping study was conducted. The study was conducted according to the method described by Sleeswijk Visser et al. [9]. A group session with four participants and semi-structured interviews with four family carers were set up. The participants were sensitized using a booklet, and then participated in a generative session. The recordings from the session and the materials made by participants were analyzed, and used as a start for the design phase.

The analysis of the generative session was based on the grounded theory, where the potential indicators of a phenomenon are discovered during the analyses [3]. The sessions were transcribed. Interesting quotes regarding current situation, information need, worries, and wishes for the future, were selected. To translate these quotes into interpreted facts, statement cards were used; cards that include a quote from the research and an interpretation of this quote. All of the statements were grouped into clusters according to their topic. The clusters were categorized into a framework to explore their relations.

Current situation: To describe the current family care situation, the quotes from the generative session were visualized using a mindmap. A number of topics described in related literature reappear in these clusters. Clusters 3 and 5 do provide new insights. Cluster 3 ('Together you know more') relates to the need for communication to get a grip on the situation and finding confirmation. Cluster 5 ('Keep an eye on the situation') represents the need for up-to-date information. The family carer can be worried at any moment of the day. Having up-to-date information available can not only support the family carer in planning and coordination, but can also provide peace of mind. Table 1 shows the clusters resulting from the study that characterize the current family care situation.

Table 1 Key clusters characterizing the present care situation.

(1) Personalized care: care activities are linked to the needs of an individual, therefore the information needs of the caregivers vary per casus and in time.
(2) Caring is done together: seniors are generally supported by multiple family caregivers who collaborate in providing care.
(3) Together you know more: towards understanding the actual care needs, caregivers need to communicate the pieces of information they collect.
(4) Include the senior: input from the senior is valuable in building awareness.
(5) Keep an eye on the situation: peace of mind is improved by updates from people visiting the senior.

Worries: Family carers get worried when things go wrong. Minor issues are often regarded as the start of bigger problems. Especially when caregivers have to judge issues from a distant location, minor issues can result in huge worries. Family carers can be worried at any place and at any time. Information is needed to take away the worries. Table 2 shows the clusters characterizing the worries.

Table 2 Overview of worries as described by the participating family carers.

(1) Love and belonging needs, including emotional state and loneliness
(2) Safety needs, including state of the home, incidents, safety, and financial needs
(3) Physiological needs, including worries about nutrition (moments of eating, preparing food), medicine, sleep (amount, moments), daily care and acute physical problems

Awareness/information needs: The participating family carers expressed the need to be better aware of the actual needs of the elder. This information need originates from their personal bond, which results in a feeling of responsibility.

Peace of mind: Lack of awareness of the situation results in worries, and consequently in a lack of peace of mind. The participating family carers expressed that need to feel up-to-date in order to feel peace of mind.

4 Design

The literature study showed that existing systems can be categorized in 4 clusters according to two main design dimensions: (1) when and how to communicate (at home or mobile), and (2) how to present sensor data (current state vs. patterns in time). As a first design step, a provocative concept was developed for each cluster (Figure 1), and the concepts were discussed with family carers.

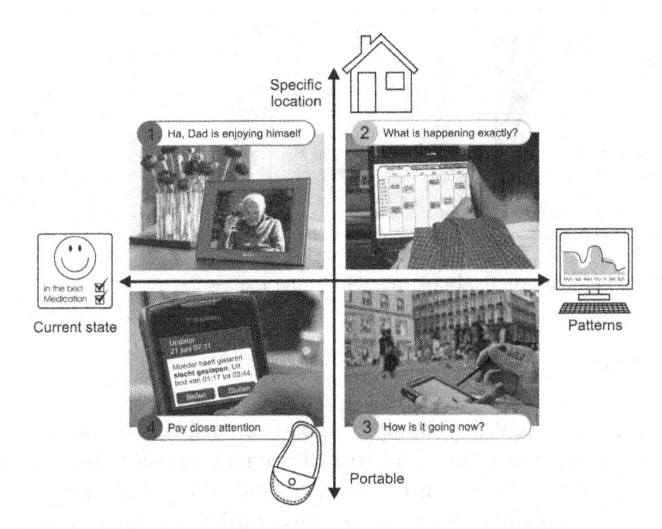

Fig. 1 Provocative design concepts were developed to trigger family carers to reflect upon their needs and the role of ICT.

The carers expressed a clear preference for a portable system, which allows them to check how the elder is doing any place any time. Their primary interest is to see how the senior is doing at a specific point in time; the family carers want to know if extra help is needed. At the same time, however, they would like to be able to see patterns in order to be able to detect changes in time.

In the interviews, the caregivers indicated that they lacked peace of mind. According to them, they wanted to know not only about the context (e.g. presence), but also about how the elderly was doing. We therefore decided to create *Family Carebook,* which combines message-based communication ('how do you feel') with monitoring ('are you at home?'). The storyboard on the next page (figure 2) describes the product-user interaction in context and over time. The application is designed for a touch screen smart phone, which enables family carers to access status information any place any time (more details will be given in sections 5).

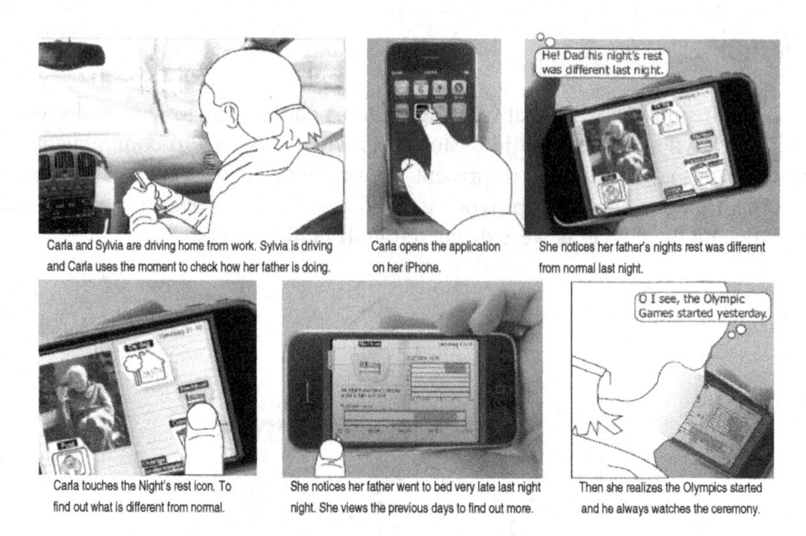

Fig. 2 Storyboard explaining the book metaphor interaction

5 Evaluation

To find out to what extent the Family Carebook contributes to the awareness and peace of mind of family carers, and to better understand the user experience, a working prototype was built. An explorative field trial was conducted in which target users could experience the concept. Two participant couples were recruited based on age and care situation. Each couple consisted of an independently living senior (age >80), and a family member or friend who regularly helps the senior.

Prototype. An iPhone application was developed for the carers. The prototype combines message-based communication, to support the social awareness, and monitoring based on sensors to provide awareness of the senior's night's rest, medication and presence. A pressure sensor was placed in the senior's bed, an infrared sensor was placed in the living room and the medication box was placed on a reed sensor to detect when it was lifted. The senior was provided with a touch screen for sending messages and viewing the sensor data.

Test set up. The field trial consisted of three parts.

Part 1 (3 days): baseline. The participants were given a booklet with assignments. The booklet was used to measure the perceived awareness level and perceived peace of mind before using the Family Carebook. Information on sleeping and medication activities was collected and used as input for the prototype.

Part 2 (5 days): prototype testing. The photo frame display and the sensors were placed in the house of the senior. The family carer was provided with the iPhone with the Family Carebook app. The supervisor explained the product use to the participants. Assignments to look at the information and send a message were

given for the first two days; the assignments helped the participants to become familiar with the prototype. After two days they could use it in their own way.
Part 3: exit interviews. An exit interview was conducted at the end of the trial.

6 Findings

Even though the field trial is limited in size and duration, the qualitative feedback can be used as an 'explorative' validation of the design choices. Participants were generally positive about their experiences with the Family Carebook, both in terms of peace of mind and of awareness. During the 8 days of the trial, the family carers kept on using the Family Carebook. The application had become part of their daily rituals. One caregiver would for example send a message to his mother every morning in order to keep up communication and to check if everything is fine. All participants would have liked to keep using the system and would have liked to introduce it to family members living further away.

Peace of mind. Whereas both couples experienced only few worries at the time of the field trial, the participating family carers did anticipate a future situation in which they would have worries. Family Carebook was considered to be a valuable addition to existing communication mechanisms. The value of the system was illustrated by the following situation: The weather was terrible outside and the family carer knew that his mother was away from her home. Without the system, the family carer would use the phone several times to check if his mother would be home safe. Using the Family Carebook, he noticed when she arrived home, he was soothed and he sent a message to ask how her evening was.

Awareness. Both participating caregivers used the Family Carebook as an additional source of information. They indicated that the Family Carebook would be especially useful for children who live at a distance from their independently living parent. The need for information was found to be different for couple B as compared to couple A. Family carer B was interested to know detailed information about the current situation. He would prefer to know more when there was little activity since this would lead to concerns; the senior could be away from home, be sitting in a chair enjoying a TV show or might have fallen down and not be capable of getting up. Therefore the sensor data resulted in extra communication (through the system or by phone). Family carer A, on the other hand, was happy with the data available through the system.

Privacy. The seniors accepted the lack of privacy caused by the sensors, since they valued the peace of mind provided by the system. Also the fact that they themselves could see what was being monitored made them part of the system, rather than being a passive subject. The messaging function also helped include the senior and provided a clear use the system. One senior often checked the Family Carebook to see how it captured her daily routines.

7 Conclusions

This paper presented a case study of a communication tool targeted at family carers of an independently living senior. A mobile phone application was developed which was aimed at supporting awareness and promoting peace of mind through sharing practical information combined with social communication.

Even though the field trial is limited in size and duration, the qualitative feedback can be used as an 'explorative' validation of the design choices. Participants did report improved peace of mind, which does suggest that a monitoring and communication tool can be used not only to improve efficiency and increase connectedness, but also to improve peace of mind.

A key limitation of the present system was found to be the 1-to-1 communication setting. Many seniors are supported by a group of caregivers, which would require group communication mechanisms. These group communication mechanisms will be studied in a next design iteration.

References

1. Birnholtz, J., Jones-Rounds: McK: Independence and Interaction: Understanding Seniors' Privacy and Awareness Needs For Aging in Place. In: CHI 2010: Privacy Awareness and Attitudes, Atlanta, GA, USA, April 10-15 (2010)
2. Consolvo, S., Roessler, P., Shelton, B.E.: The CareNet Display: Lessons Learned from an In Home Evaluation of an Ambient Display. In: Davies, N., Mynatt, E.D., Siio, I. (eds.) UbiComp 2004. LNCS, vol. 3205, pp. 1–17. Springer, Heidelberg (2004)
3. Glaser, B.G., Strauss, A.L.: The discovery of grounded theory: Strategies for qualitative research. Aldine Publishing, Chicago (1967)
4. Khan, V.J., Markopoulos, P., Eggen, B., Metaxas, G.: Evaluation of a pervasive awareness system designed for busy parents. Pervasive and Mobile Computing (2010)
5. Khan, V.J., Markopoulos, P., Mota, S., IJsselsteijn, W., de Ruyter, B.: Keeping in Touch with the Family: Home and Away with the ASTRA Awareness System. In: CHI 2004 (2004)
6. Metaxas, G., Metin, B., Schneider, J., Markopoulos, P., de Ruyter, B.: Awareness of Daily Life Activities. In: Awareness Systems Advances in Theory, Methodology and Design, pp. 351–365 (2009)
7. Mynatt, E.D., Rowan, J., Jacobs, A., Craighill, S.: Digital Family Portraits: Supporting Peace of Mind for Extended Family Members. In: CHI 2001 (2001)
8. Oulasvirta, A., Raento, M., Tiitta, S.: ContextContacts: Re-designing smartphone's contact book to support mobile awareness and collaboration. In: Proceedings of Mobile HCI 2005, pp. 167–174. ACM Press, New York (2005)
9. Sleeswijk Visser, F., Stappers, P.J., van der Lugt, R., Sanders, E.B.-N.: Contextmapping: Experiences from practice. Codesign 1(2), 119–149 (2005)
10. Vastenburg, M.H., Vroegindeweij, R.J.: Designing an Awareness Display for Senior Home Care Professionals. In: Tscheligi, M., de Ruyter, B., Markopoulus, P., Wichert, R., Mirlacher, T., Meschterjakov, A., Reitberger, W. (eds.) AmI 2009. LNCS, vol. 5859, pp. 215–224. Springer, Heidelberg (2009)

Stress Monitoring in Conflict Resolution Situations

Davide Carneiro, José Carlos Castillo Montotya, Paulo Novais,
Antonio Fernández-Caballero, José Neves, and María Teresa López Bonal

Abstract. Online Dispute Resolution is steadily growing to become the major alternative to litigation in court. In fact, given the characteristics of current disputes, technology-based conflict resolution may be a quite efficient approach. However, in this shift of paradigm, there are also threats that should be considered. Specifically, in this paper we deal with the problem of the lack of important context information when parties communicate in a virtual setting. In that sense, we propose the addition of a monitoring framework capable of measuring the level of stress of the parties in a non-invasive way. This information will be used by the platform and the mediator throughout the complete conflict resolution process to adapt strategies in real-time, resulting in a context-aware and more efficient approach.

Keywords: Online Dispute Resolution, Stress Monitoring, Monitoring Framework.

1 Introduction

The field of conflict resolution is historically related to the Social Sciences field. However, the advent of technology brought along significant changes in the way that conflicts emerge and are solved [11]. This is especially visible in electronic contracting, in which the soundest example is e-commerce. This has led to the emergence of the so-called Online Dispute Resolution (ODR) [1], encompassing

Davide Carneiro · Paulo Novais · José Neves
Department of Informatics/CCTC,
University of Minho, Braga, Portugal
e-mail: {dcarneiro,pjon,jneves}@di.uminho.pt

José Carlos Castillo Montotya · Antonio Fernández-Caballero · María Teresa López Bonal
Instituto de Investigación en Informática de Albacete (I3A),
Universidad de Castilla-La Mancha, Albacete, Spain
e-mail: {JoseCarlos.Castillo,Antonio.Fdez,Maria.LBonal}@uclm.es

P. Novais et al. (Eds.): Ambient Intelligence - Software and Applications, AISC 153, pp. 137–144.
springerlink.com

techniques or approaches for conflict resolution taking partly or entirely place in a virtual environment, with a varying autonomy of technology [2].

Nevertheless, ODR also presents some threats [3]. One of the most significant is the important context information that is lost when parties do not communicate in the physical presence of each other. Particularly significant is all the information from the body language that we (unconsciously) rely on in our day-to-day interactions. In fact, Mehrabian [4] states that the non-verbal elements of a conversation are particularly important for communicating feelings and attitudes. Thus, the problem is that information is lost in a virtual setting and makes it hard for the intervenient parties to understand the emotional state of each other.

In this paper we describe a context-aware environment to support a conflict resolution platform by providing important information about the state of the parties [5]. Specifically, we focus on the estimation of the level of stress of the intervenient. The aim is to improve the awareness of user's states in current ODR processes, allowing both parties and neutrals to take better and more informed decisions, ultimately increasing the efficiency of the conflict resolution process.

2 System Overview

Information about the user's context is provided through a monitoring framework [7], which is customized to perform movement detection from a camera located in front of the user. Considering frame-to-frame movement information, the level of agitation or excitement of the user can be analyzed. From these movements, more complex behaviors can be analyzed to detect stress patterns in the form of activities [10]. Moreover, recent works deal with the detection of emotions by employing visual sensors [9]. Additionally, we are relying on the use of interfaces developed for touch screens empowered with accelerometers that allow capturing information about how the user interacts with the conflict resolution system. Specifically, we are interested in information about the intensity and the accuracy of the touches, basically the users touch patterns. Combining this information, an estimation of the level of stress of a user can be computed and provided to the conflict resolution platform (Figure 1).

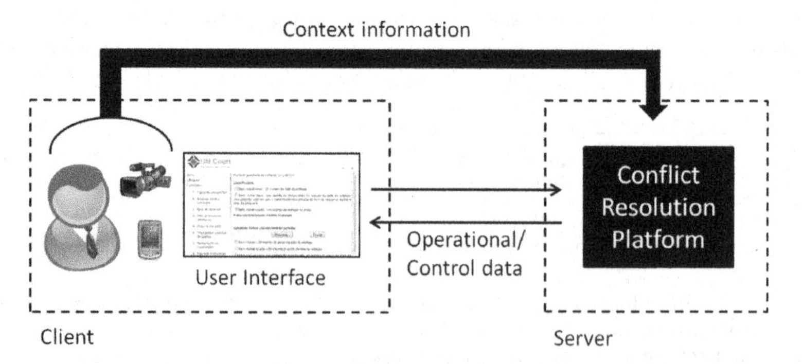

Fig. 1 Overview of the developed system: context information is provided to the conflict resolution platform, which interpret the user's state.

Information about stress is very important in a conflict resolution scenario, as it will allow the mediator or the platform itself to adapt to significant changes in the user state. Concrete adaptations include temporarily interrupting the direct contact between the parties when they become too stressed, making a pause in the process, or even proposing an outcome that satisfies more precisely the expectations of the more stressed party, in an attempt to calm him/her down and keep him/her focused on the process. In the following sections we describe in detail the approach used, the experiments performed and the results achieved.

3 Monitoring the User Stress from Visual Information

To detect the excitement degree of a user, the framework recently presented in [7] has been particularized to monitor the movements of a user seated in front of a camera while interacting with the ODR touch interface (Figure 2). After data acquisition, a timestamp is associated to the images to enable them to be matched with the touch information. The amount of movement is calculated as the amount of pixels that have changed their RGB color value over a given threshold between two consecutive images. Although the algorithm is deeply described in [8], including an implementation focused on the infrared imaging technology, the main steps are highlighted next. Firstly the subtraction, I_S, of the current image, I_t, and the previous one, I_{t-1}, is calculated as follows:

$$I_S(i,j) = |I_t(i,j) - I_{t-1}(i,j)|$$

Afterwards, it is necessary to determine which pixels in I_S have higher changes, just in order to discard noise. That is why a binarization process is carried out. Thus, pixels over the established threshold, τ, are set to 1 and those pixels below τ are set to 0 as described next , obtaining the binarized image, B.

$$B(i,j) = \begin{cases} 0, & I_S(i,j) < \tau \\ 1, & I_S(i,j) \geq \tau \end{cases}$$

After the binarization, some filters are applied to B in order to reduce the noise impact and to enhance the really relevant movements. For this purpose, some opening and closing morphological operations are applied, obtaining B^F. Finally, the amount of movement is calculated as the number of non-zero pixels on B^F:

$$M = \sum_{j=0}^{height} \sum_{i=0}^{width} B^F(i,j)$$

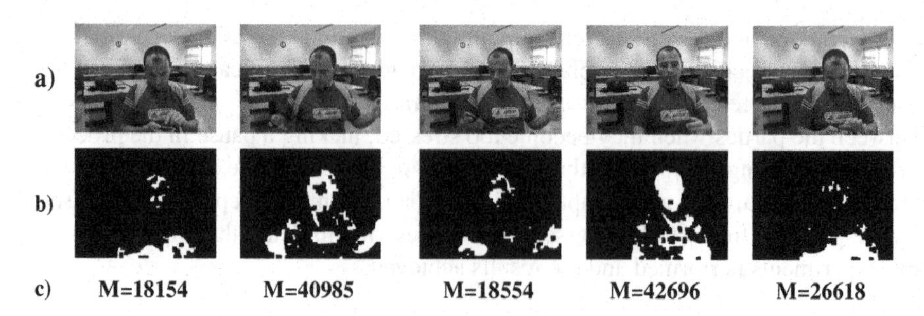

Fig. 2 Results of movement detection of a user whilst interacting with the conflict resolution platform. Row a) shows input images, row b) shows binarized and filtered movement and row c) shows the amount of movement detected.

4 Collecting Information from User Interaction

The effects of stress are generally visible in almost all of our behaviors. In that sense, it is possible to establish a relation between the effects observed and the level of stress. Specifically, we are interested in how stress affects the user's touch patterns while using the interface of the application. In that sense, the proposed system constantly collects information about the user's touch intensity and accuracy. The approach is based on the notion that a higher level of stress results in a higher touch pressure and a lower accuracy and vice versa. The information collected is sent to the UMCourt conflict resolution platform [6], which interprets it and uses it as an additional input for managing the conflict resolution process.

The data is always analyzed taking into consideration the context in which it was acquired. In that sense, data generated from the management (e.g. managing personal information, accessing past cases) and login interfaces are used to calibrate the system and to establish a baseline for the level of stress. This comprises the training phase. On the other hand, data generated from the current conflict resolution interfaces are used to assess the level of stress of the user. It constitutes the operational phase. This is important as two individuals are in no case affected by the stress in the same way. In order to compute a measure of accuracy, touch interfaces have two areas: active and passive (see Figure 3). Active areas are defined by active controls used by the user to perform his/her tasks. Passive areas have no controls and there is no reason why a user should touch them. This allows measuring the accuracy of the touch by comparing the total number of touches versus the touches in active or passive areas.

To measure the intensity of the touch, the whole touch action is analyzed, since the finger of the user firstly touches the screen and then he/she releases it. In fact, differences between a touch performed by a stressed user and by a user in a normal state are easily visible (as shown in Figure 4). In this work, it can be concluded that, in general, the touch pattern of a stressed user has a longer duration and starts with an increasing pressure up to a maximum reached and then it decreases until the release of the finger. On the other hand, the touch pattern of a user in a normal state usually starts in or near a maximum value of pressure, which then decreases until the finger is released.

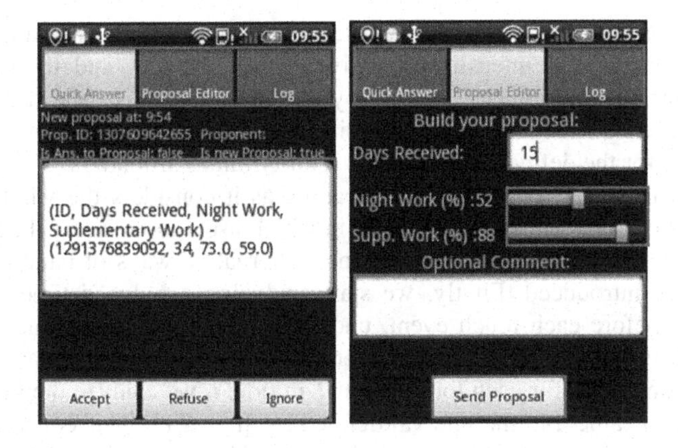

Fig. 3 A detail of the touch user interfaces used in the experiment: colored rectangles define active areas while the rest are passive areas.

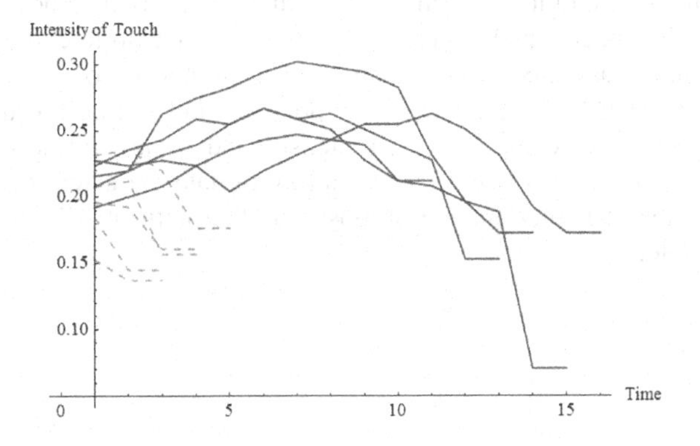

Fig. 4 Plot of the variation of the touch intensity over time from touch events generated by a user in a normal state (dashed orange line) and in a stressed state (blue line).

5 Experiments and Results

To perform the experiments detailed in this paper, we place the users of the conflict resolution platform in front of video cameras, while they interact with the touch interfaces. The data collected from the touch interfaces consist of a list of touch events. Each entry includes the timestamp in which it occurs, the maximum, minimum and mean value of the touch intensity as well as the target of the touch, i.e. passive or active. As stated before, the data collected from the camera consists of a sequence of images, captured from a frontal view of the user. The images are associated to a timestamp to allow the correspondence with the touch events. As an output, the amount of movement from the color input images is obtained.

Given that the visual system generates much more information than the touch interfaces, a recurrent task is to group parts of the information from the camera. The first analysis of the information starts by grouping all the information from the camera generated between two consecutive touch events T_{t-1} and T_t. This allows to compare the intensity of a given touch event T and the amount of movement in the time interval defined by $]t-1, t]$. For this data, the Pearson's correlation coefficient between the intensity of the touch and the sum of the ratio of movement for the defined time interval returns values around 0.5.

However, this approach may be misleading as it considers the whole interval between touch events and there are cases in which users move away their focus of attention from the application. In that sense, alternative ways of handling the information are introduced. Firstly, we start analyzing a reduced time interval Δ immediately before each touch event, under the conviction that immediately before interacting with the interface, the user's attention is directed towards the conflict resolution. Thus, for each touch event T in time t, we analyze the information from the video camera in the interval defined by $[\![t-\Delta, t]\!]$. This consists in computing the sum of the movement in each interval and comparing it with the intensity of the touch. The experiment shows a Pearson's correlation of around 0.68.

We also analyze a concrete interval, which is the most significant in terms of stress in a conflict resolution system: an interval after receiving a new proposal. In fact, when parties receive and read a new proposal, they tend to express in a more explicit way their frustration or their contentment. In that sense, when considering only intervals of data starting in each new proposal and lasting for an interval Δ, the highest value of correlation between intensity and amount of movement is visible, namely around 0.74. Figure 5 shows a plot of data extracted from a negotiation session where 10 solutions are proposed by the platform. The correlation of the data is visible.

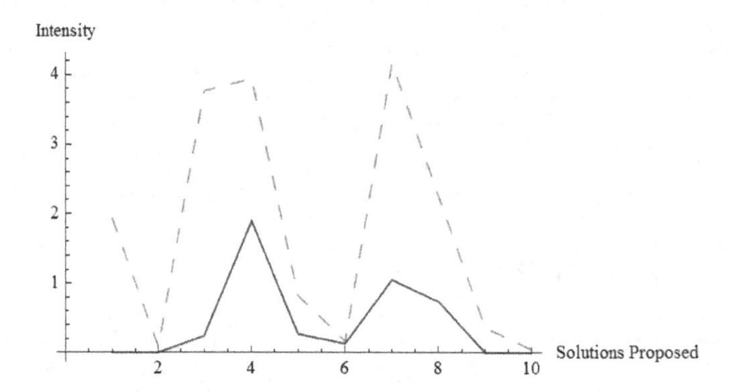

Fig. 5 Plot of data concerning the sum of the intensity of the touches (dashed plot) and the mean rate of movement after the platform proposes a solution.

Two other representations of the collected data allow seeing this relation between increased amount of movement and intensity of the touch in stressed users. In the first one (see Figure 6a)), the intensity of the touch versus the sum of the movement in the period after the last touch and before the current one is analyzed. It is possible to conclude that, in a general way, the touches with higher intensity are preceded by higher movement rates. The second representation of the data depicts the movement immediately before touch events (see Figure 6 b)). Here, it is also possible to see that touch events with smaller intensity are associated to less movement, and vice versa.

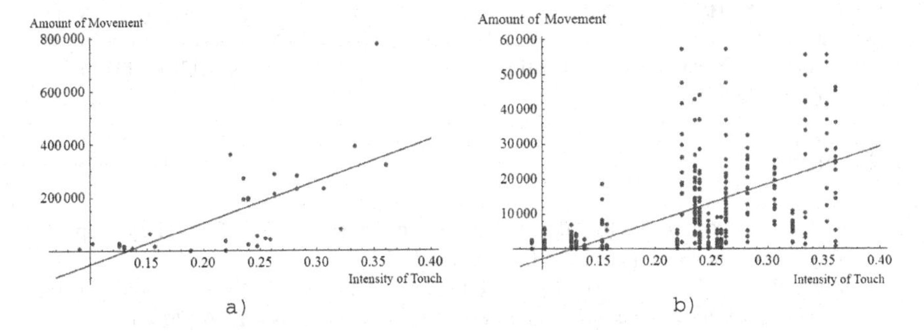

Fig. 6 Two different representations of the data, highlighting the relation between the amount of movement and the intensity of the touch.

6 Conclusions and Future Work

After performing the experiments described in this paper, we are able to conclude that effectively there is a relation between increased user movement and touch intensity and accuracy. Even though, the amount of movement is a first step to the visual detection of the user stress that can be improved by adding more complex techniques, such as methods based in pattern recognition to detect specific movements of the user. Nevertheless, user movement has been proved to be a good stress indicator in combination to other sensors information, such as the touch. Moreover, it can also be concluded that the relation movement-touch is more emphasize in moments of more stress, e.g. after parties receive disadvantageous proposals. In that sense, we can assume that there is a relation between higher levels of stress and higher touch intensities and movements. The approach can thus be an effective and non-invasive way of providing important context information to the conflict resolution platform and to the mediator. By taking in account this information it is possible to build context aware platforms capable of adapting in real-time to the state of the users, aiming for an increased success rate.

This work goes on by including new sources of information that allow a more accurate classification of the level of stress. Specifically, we are considering including information from accelerometers available in handheld devices used for user interaction. This will help in finding patterns associated to stress. We are also considering the use of additional technologies, namely electroencephalograms. Moreover, regarding the information provided by the monitoring framework, the current work focuses on the identification of human gestures to detect patterns or activities. Also, the identification of simple emotions will be added to extend the user mood detection to a wider range of human feelings.

Acknowledgments. The work described in this paper is included in TIARAC - *Telematics and Artificial Intelligence in Alternative Conflict Resolution Project* (PTDC/JUR/71354/2006), which is a research project supported by FCT (Science & Technology Foundation), Portugal. The work of Davide Carneiro is supported by a doctoral grant by FCT (SFRH/BD/64890/2009). This work is also partially supported by the Spanish Ministerio de Ciencia e Innovación under project TIN2010-20845-C03-01 and by the Spanish Junta de Comunidades de Castilla-La Mancha under projects PII2I09-0069-0994 and PEII09-0054-9581.

References

1. Katsch, E., Rifkin, J.: Online dispute resolution – resolving conflicts in cyberspace. Jossey-Bass Wiley Company, San Francisco (2001)
2. Peruginelli, G., Chiti, G.: Artificial Intelligence in alternative dispute resolution. In: Proceedings of the Workshop on the Law of Electronic Agents – LEA (2002)
3. Larson, D.: Technology Mediated Dispute Resolution. In: Proceeding of the 2007 Conference on Legal Knowledge and Information Systems: JURIX 2007: The Twentieth Annual Conference. IOS Press, Amsterdam (2007)
4. Mehrabian, A.: Silent Messages – A Wealth of Information about Nonverbal Communication. Personality & Emotion Tests & Software (2009)
5. Carneiro, D., Novais, P., Neves, J.: Toward Seamless Environments for Dispute Prevention and Resolution. In: Novais, P., Preuveneers, D., Corchado, J.M. (eds.) ISAmI 2011. AISC, vol. 92, pp. 25–32. Springer, Heidelberg (2011), doi:10.1007/978-3-642-19937-0_4
6. Carneiro, D., Novais, P., Neves, J.: Towards Domain-Independent Conflict Resolution Tools. In: The 2011 IEEE/WIC/ACM International Conference on Intelligent Agent Technology, IAT 2011 (2011)
7. Castillo, J.C., Rivas Casado, A., Fernández-Caballero, A., López, M.T., Martínez-Tomás, R.: A Multisensory Monitoring and Interpretation Framework Based on the Model-View-Controller Paradigm. IWINAC (1), 441–450 (2011)
8. Fernández-Caballero, A., Castillo, J.C., Martínez-Cantos, J., Martínez-Tomás, R.: Optical flow or image subtraction in human detection from infrared camera on mobile robot. Robotics and Autonomous Systems 58(12) (2010) ISSN 0921-8890
9. Bee Theng, L.: Portable real time emotion detection system for the disabled. Expert Systems with Applications 37(9), 6561–6566 (2010), j.eswa.2010.02.130
10. Fernández-Caballero, A., Castillo, J.C., Rodríguez-Sánchez, J.M.: A Proposal for Local and Global Human Activities Identification. In: Perales, F.J., Fisher, R.B. (eds.) AMDO 2010. LNCS, vol. 6169, pp. 78–87. Springer, Heidelberg (2010)
11. Katsh, E., Rifkin, J., Gaitenby, A.: E-Commerce, E-Disputes, and E-Dispute Resolution: In the Shadow of eBay Law. Ohio State Journal on Dispute Resolution 15, 705 (1999)

Discovering Mobility Patterns on Bicycle-Based Public Transportation System by Using Probabilistic Topic Models

Raul Montoliu

Abstract. In this work, we present a new framework to discover the daily mobility routines which are contained in a real-life dataset collected from a bike-sharing system. Our goal is the discovery and analysis of mobility patterns which characterize the behavior of the stations of a bike-sharing system based on the number of available bikes along a day. An unsupervised methodology based on probabilistic topic models has been used to achieve these goals. Topic models are probabilistic generative models for documents that identify the latent structure that underlies a set of words. In particular, Latent Dirichlet allocation (LDA) has been used to discover mobility patterns. Our database has been collected for almost half a year from the *Bicicas* bike sharing system in Castellón (Spain). A set of experiments have been conducted to demonstrate the type of patterns that can be effectively discovered by using the proposed framework.

1 Introduction

Public bike sharing services are becoming more and more popular in the last years. In general, the basic premise of the bike sharing concept is sustainable transportation, so, they have been introduced to improve air quality, to reduce traffic congestion and to increase mobility choices in crowded cities. Recently, it has been estimated that, at the end of 2010, there were more than 200 such schemes operating worldwide [2].

Public bike sharing services work as follows: users can pick up a bike in some of the stations located over the city. Then, they ride the bike to move to their destination. Finally, they return the bike into another station (or into the same) before the rental period finishes. Two are the biggest problems detected: on the one hand,

Raul Montoliu
Institute of New Imaging Technologies (INIT),
Jaume I University, Castellón, Spain
e-mail: montoliu@uji.es

P. Novais et al. (Eds.): Ambient Intelligence - Software and Applications, AISC 153, pp. 145–153.
springerlink.com

the impossibility to find a bike when users want to start a journey (because there are not available bikes in the departure station) and, on the other hand, the impossibility to return the bike in the destination (because the arrival station is full). There are basically two ways to solve these problems: 1) to inform users in advance about the best places to pick up or leave the bikes and 2) to improve the redistribution of bikes from full to empty stations. To allow users to plan their routes in advance, many systems provide on their website a map of stations, where users can check the status of the stations (amount of available bikes and empty slots) close to their departure and arrival points. To perform the bikes redistribution task, in many systems there are trucks which move bicycles from highly loaded stations to empty ones.

In this paper, a bike sharing system of a mid-size city has been studied in order to discover mobility patterns by using probabilistic topic models. Our main objective is to discover mobility patterns as a preliminary step in the goal of a better understanding of the behavior of the bike-sharing systems. On the one hand, thanks to the discovery of the mobility patterns, it may be possible (in a future work) to improve the route planning of the trucks, a fact that could result in a better performance of the whole system improving the users' satisfaction level. On the other hand, the resulting mobility patterns could help to the system designers in the task of determining the location of future stations and also to define the system parameters to improve its performance.

In this work, probabilistic topic models have been used to discover mobility patters in a unsupervised manner. In particular, Latent Dirichlet Allocation (LDA) [1] has been used. Topic models are generative models that represent documents as mixtures of topics, learned in a latent space. They are advantageous to activity modeling tasks due to their ability to effectively characterize discrete data represented by bags (i.e. histograms of discrete items). These models can capture which words are important to a topic as well as the prevalence of those topics within a document. It has been showed previously [3], that topic models prove to be effective in making sense of behavioral patterns at large-scale while filtering out the immense amount of noise in real-life data, as it is the case of our data collected from bike sharing systems.

The mobility data have been collected from the web page of a bike-sharing system of a mid-size city and consist on the number of available bikes in all stations every five minutes during the opening hours of the system. Although there are some previous works using mobility data collected from a bike sharing system [4, 5, 7], to our knowledge, this is the first study, using this kind of mobility data, that tries to discover mobility patterns by using topic models.

The main contributions of this work are the following ones:

1. We discover the mobility patterns of a mid-size city from mobility data obtained from a bike sharing system.
2. We devise a novel bag representation of a day in a bike sharing station which capture the behavior of the station in specific time moments.
3. We propose a methodology for the automatic discovery of daily routine patterns with LDA.

This paper has been organized as follows: Section 2 discusses the related work. An overview of our work is outlined in Section 3. Section 4 explains our representation methodology. We present and discuss the experimental results in Section 5. Finally, Section 6 draws the main conclusions arisen from this work.

2 Related work

In [4], an analysis of the temporal and spatial patterns of the *Bicing* bike-sharing system (Barcelona, Spain) was conducted in order to explore the human behavior and movements dynamics in the city of Barcelona. They showed that people daily routines, culture, and station location have a big influence on stations usage patterns. They also performed a clustering technique to classify the stations in different types according to station usage. In a subsequent work of the same authors [5], they focused on predicting the number of available bicycles at each station at a given time in the future. In [7], similar studies have been conducted to predict the future usage of the stations and to cluster stations according to their usage. The main differences between these approaches are in the specific techniques used to cluster data and to predict the future state of the stations.

On the other hand, topics models are powerful tools initially developed to characterize text documents, but in the last years they have been extended to other collection of discrete data in several different problems, e.g. to discover human routines from mobile phone data [3], to perform trajectory analysis and semantic region modeling on video scenes [10], for scene categorization and object recognition [9], and human action recognition [8], among others. They are probabilistic generative models that can be used to explain multinomial observations by unsupervised learning. One of the most popular topic model is Latent Direchlet Allocaltion (LDA). It is a generative model, introduced by Blei et al. in [1], in which each document is modeled as a multinomial distribution of topics and each topic is modeled as a multinomial distribution of words. Several approximation techniques have been developed for inference and learning in the LDA model [1, 6]. In this work we adopt the Gibbs sampling approach [6].

3 Overview of Our Work

Figure 1 shows the framework of our approach. Data has been collected from the web page of a bike sharing system. We collected the available number of bikes at each station every five minutes during the opening time period of the system. Raw data has been transformed into the bag of words representation (i.e. the document database) following the procedure explained in Section 4. Then, LDA procedure is applied to discover mobility routines.

Our data has been collected from the *Bicicas*[1] bike-sharing system in Castellón (Spain). Castellón is a mid-size city of almost 200.000 inhabitants. There are 45

[1] http://www.bicicas.es/

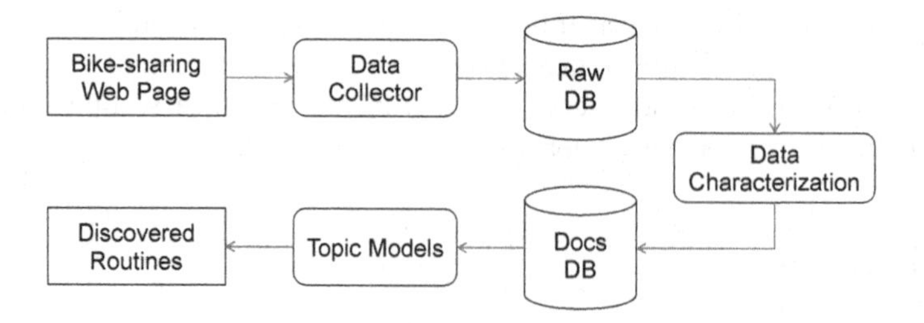

Fig. 1 Framework of the approach presented in this paper.

stations distributed throughout the whole city. Nowadays, there are almost 20.000 users and the average number of trips per day is around 3.800. *Bicicas* works from 7:30 AM to 10:30 PM on workdays, from 9 AM to 9 PM on Saturday, from 9 AM to 8 PM on summer Sundays and from 9 AM to 6 PM on the rest of Sundays. After having picked up a bike, a user is allowed to ride the bike for up to two hours. Data has been collected for almost half a year, starting on April, 13th and finishing on September, 30th. In total, almost 1.5 million of data elements have been collected.

4 Bag of Words Representation

In LDA terminology, the entity termed *word* is the basic unit of discrete data defined to be an item from a vocabulary of size V. A *document* is a sequence of words. A *corpus* is a collection of M documents. There are K latent topics in the model, where K is defined by the user. The main objectives of LDA inference are (see [1, 6] for details): 1) to find the probability of the l-th word of the vocabulary ($l = 1,\ldots,V$) given the k-th topic estimated ($k = 1,\ldots,K$), $p(w_l|z_k)$, and 2) to find the probability the k-th topic estimated given the m-th document of the corpus, $p(z_k|d_m)$.

Our collected data is a corpus of M documents: $\Pi = \{d_1,\ldots,d_M\}$, where each document d_m represents a day in a particular station using a vector of N_m data elements: $d_m = [\phi_1^m,\ldots,\phi_{N_m}^m]$. Each data element ϕ_i^m is the number of available bikes for a particular station and a particular date (i.e., the m-th document) order by time. Note that not all documents have the same number of elements, since opening hours are not the same in all days.

In order to capture the behavior of the stations in different moments of the day, each day is divided into 8 time periods: $\{P_7,P_9,P_{11},P_{13},P_{15},P_{17},P_{19},P_{21}\}$. Each time period represents a particular moment in a day: P_7, early morning (7:30-9 AM); P_9, mid-morning (9-11 AM); P_{11}, late morning (11 AM-1 PM); P_{13}, lunch time (1-3 PM); P_{15}, afternoon (3-5 PM); P_{17}, late afternoon (5-7 PM); P_{19}, early evening (7-9 PM); and P_{21}, late evening (9-10:30 PM).

In order to characterize a time period, the difference between the number of available bikes in the end of the period with respect to the beginning is calculated. When this difference is positive and bigger than a threshold τ, the period is encoded with the symbol $\nearrow$ (increment). When the difference is negative and its absolute value

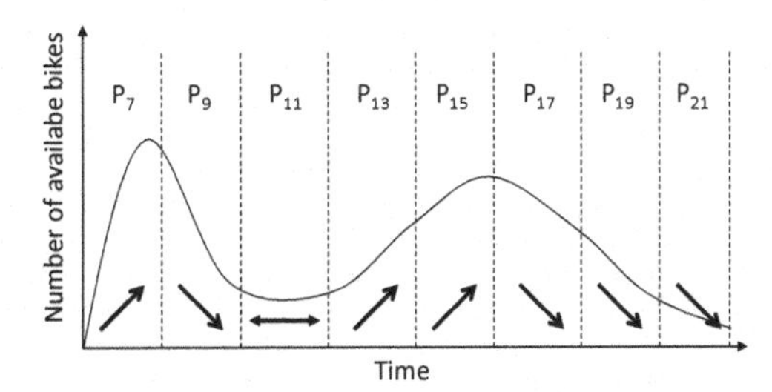

Fig. 2 Number of available bikes of a hypothetic document. Vertical dashed lines are the period divisions. The bottom arrows show the behavior of the station in each period.

is bigger than τ, it is encoded with $\searrow$ (decrement). Finally, if the absolute value of the difference is lower than τ then we assume that there is not a significant change in the number of bikes. In this case, the period is encoded with $\leftrightarrow$ (no change). The bag of word of each document d_m consists in a set of 8 words, one for each time period. A word is a combination of the time period and the behavior of the station in this period. We name each word as $P_A B$, where P_A represents the time period, $A \in \{7,9,11,13,15,17,19,21\}$, and B the behavior, $B \in \{\nearrow, \searrow, \leftrightarrow\}$. In all experiments reported in this paper, $\tau = 2$.

Figure 2 shows an illustrative example of how the bag of words for a particular day and station are encoded. It shows the number of available bikes of an hypothetic document. The bottom arrows show the behavior of the station in each period. In this example, the words extracted for this day are as follows: '$P_7 \nearrow$', '$P_9 \searrow$', '$P_{11} \leftrightarrow$', '$P_{13} \nearrow$', '$P_{15} \nearrow$', '$P_{17} \searrow$', '$P_{19} \searrow$' and '$P_{21} \searrow$'.

5 Experiments and Results

One of the main limitation of LDA is that the number of latent topics must to be set a priori. Perplexity has been used as a measure to determine automatically the optimal number of latent topics K. We computed perplexity for LDA using K values from 10 to 500 with increments of 10 (i.e. 50 samples). For all values of K, 1000 iteration of the Gibbs sampling algorithm have been performed. LDA parameters (α, β) have been set to $\alpha = 50/K$ and $\beta = 0.1$. The perplexity for a particular number of topics K is estimated as follows:

$$perplexity(K) = e^{entropy(K)} \tag{1}$$

$$entropy(K) = -\sum_{k=1}^{K} p(z_k) \left(\sum_{l} p(w_l|z_k) log(p(w_l|z_k)) \right) \tag{2}$$

$$p(z_k) = \sum_{m=1}^{M} p(z_k|d_m) p(d_m) \tag{3}$$

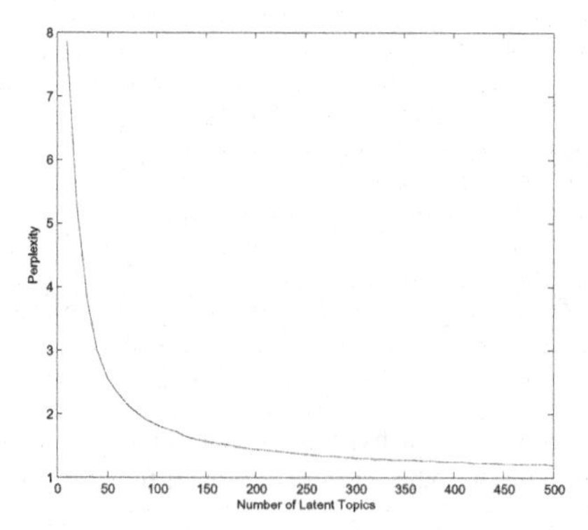

Fig. 3 Perplexity plot as a function of latent topics. After $K = 100$ perplexity starts to stabilize to a low value.

where $p(z_k)$ is the probability of k-th topic and $p(d_m)$ is probability of the m-th document. In this work, we assume that all documents are equally probables. The perplexity is plot over the number of latent topics in Figure 3. A significant drop in perplexity occurs at approximately $K = 100$ topics. Afterwards, the perplexity tends to stabilize. Thus, we choose $K = 100$ as the number of latent topics for the remaining experiments.

In order to discover mobility patterns, the LDA algorithm has been applied to our mobility data. The LDA model successfully found latent topics over all days allowing to extract the dominating mobility patterns present in the data. The discovery of mobility patterns by means of the proposed unsupervised methodology revealed different types of mobility patterns, which can be used to explain the behavior of a station in a day by following characteristic trends to various topics with a probability measure. For space limitation we can not show all the discovered patterns. To illustrate the patterns discovered for 3 representative topics, we rank the 20 most probable documents, ranked by $p(z_k|d_m)$, and visualize them in Figure 4. Some interesting results are as follows:

- **Topic 25:** Its 3 most probable words, ranked by $p(w_l|z_k)$, are: '$P_{13} \nearrow$', '$P_{15} \searrow$', and '$P_9 \nearrow$'. Topic 25 captures the pattern: *"The number of available bikes increases at lunch time and decreases after lunch"*. This can occur in stations located near places where people usually go to have lunch, as homes, restaurants, etc.
- **Topic 43:** Its 3 most probable words, ranked by $p(w_l|z_k)$, are: '$P_9 \nearrow$', '$P_{11} \searrow$', and '$P_{21} \searrow$'. Topic 43 captures the pattern *"The number of available bikes increases at early morning and decreases at late morning"*. This can occur in stations located near places where people go to work or to study and they use to go early to have lunch out of their work/study places.
- **Topic 57:** Its 3 most probable words, ranked by $p(w_l|z_k)$, are: '$P_{21} \searrow$', '$P_{19} \nearrow$', and '$P_{17} \leftrightarrow$'. Topic 57 captures the pattern *"The number of available bikes increases at early evening and decreases at late evening"*. This behavior can be

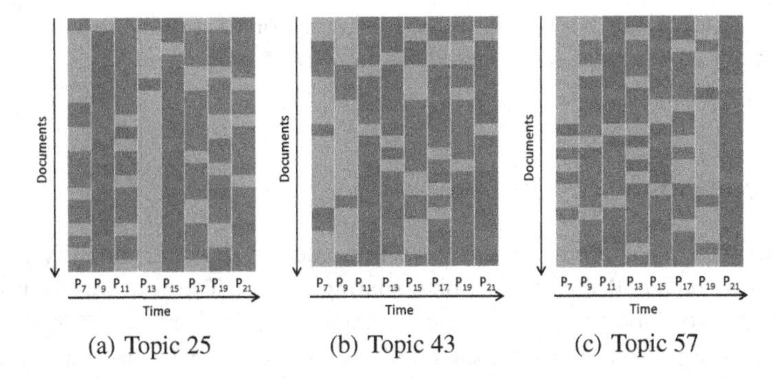

(a) Topic 25 (b) Topic 43 (c) Topic 57

Fig. 4 The 20 most probable documents, according to $p(z_k|d_m)$, for topics 25, 43 and, 57. A document is represented by a row (of 20 pixels) where each period has been colored according to its behavior. It is green when the number of available bikes increases, it is red when this number decreases and blue when the number does not change during the period. Vertical white lines show the period limits.

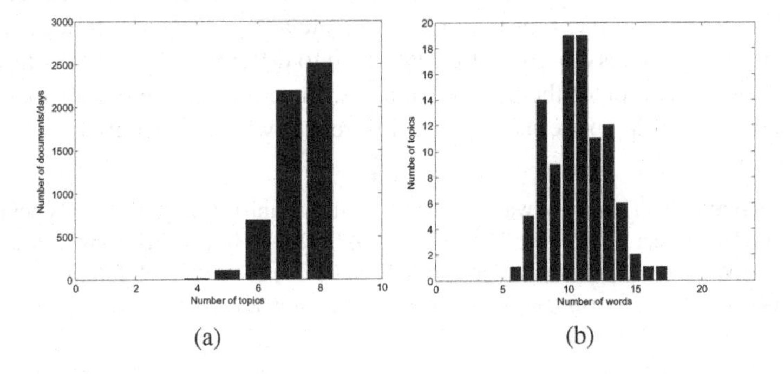

(a) (b)

Fig. 5 a) Histogram of the number of "dominating" topics per document. b) Histogram of the number of words needed to describe a topic.

explained in two situations. The first one, in stations located near places where people go after work to spend some leisure time and then, go back home. The second one, when people go back home after work. In this last situation, the emergence of the word '$P_{21} \searrow$' can be explained due to an operation of the truck that pick up bikes to redistribute them in some other stations, since at early morning, many users are going to return a bike, and therefore, it is necessary to have many free slots.

One fundamental question that arise is about how evident is the "mixture of topics" assumption in our data, i.e. could a document be "explained" just by using a single topic or more than one is needed? Similar to [3], we show in Figure 5a a histogram of the number of "dominating" topics per document. To compute the histogram, for each document m, we counted the number of topics with $p(z_k|d_m) > T_d$. In general, the majority of the documents are well described using 7 or 8 topics. We also show in 5b a histogram of the number of words needed to describe a topic, showing that, in general, a topic can be well described using 10 or 11 words. To compute the histogram, for each topic k, we counted the number of words with $p(w_l|z_k) > T_z$. In both cases, we used $T_d = T_z = 0$.

6 Conclusions

In this paper, a new framework has been presented to discover mobility patterns from mobility data collected from a bike-sharing system. The goal was to discover and to analyze of mobility patterns which characterize the behavior of the stations of a bike-sharing system based on the number of available bikes along a day. Probabilistic topic models have been used to discover mobility patterns in an unsupervised manner. In particular, Latent Dirichlet Allocation (LDA) has been used. Our experiments have proved that our method to describe a day using a novel bag-based representation together with the methodology for the automatic discovery of daily routine patterns with LDA, can effectively discover latent topics which describe mobility patterns. Future work will focus on using the topics discovered to classify the documents into several day types in order to create a bag-of-words strategy to perform a clustering process to group the stations into different types. In addition, we will focus on using type of the day information, since it might be possible that, for some stations, workday patterns could be different to weekend patterns.

Acknowledgements. This work was supported by the Spanish Ministerio de Ciencia e Innovación under project Consolider Ingenio 2010 CSD2007-00018. We thank Adriá Martí for his contribution to data collection, Henry Anaya (INIT) for his insightful comments and Benjamín Garcés (Maquiver S.L.) for providing information about the *Bicicas* system.

References

1. Blei, D.M., Ng, A.Y., Jordan, M.I.: Latent dirichlet allocation. Journal of Machine Learning Research 3, 993–1022 (2003)
2. Dearnaley, M.: Agency seeks substitute for nextbike's shelved hire fleet (2011), http://www.nzherald.co.nz/auckland-region/news/article.cfm?l_id=117&objectid=10700595 (accessed January 20, 2011)
3. Farrahi, K., Gatica-Perez, D.: Discovering routines from large-scale human locations using probabilistic topic models. ACM Transactions on Intelligent Systems and Technology 2, 3:1–3:27 (2011)
4. Froehlich, J., Neumann, J., Oliver, N.: Measuring the pulse of the city through shared bicycle programs. In: Proceedings of International Workshop on Urban, Community, and Social Applications of Networked Sensing Systems (UrbanSense 2008) (2008)
5. Froehlich, J., Neumann, J., Oliver, N.: Sensing and predicting the pulse of the city through shared bicycling. In: Proceeding of the Twenty-First International Joint Conference on Artificial Intelligence (IJCAI 2009) (2009)
6. Griffiths, T.L., Steyvers, M.: Finding scientific topics. Proceedings of the National Academy of Sciences of the United States of America 101(1), 5228–5235 (2004)

7. Kaltenbrunner, A., Meza, R., Grivolla, J., Codina, J., Banchs, R.: Urban cycles and mobility patterns: Exploring and predicting trends in a bicycle-based public transport system. Pervasive and Mobile Computing 6(4), 455–466 (2010)
8. Niebles, J., Wang, H., Fei-Fei, L.: Unsupervised learning of human action categories using spatial-temporal words. International Journal of Computer Vision 79, 299–318 (2008)
9. Sudderth, E.B., Torralba, A., Freeman, W.T., Willsky, A.S.: Describing visual scenes using transformed objects and parts. International Journal on Computer Vision 77, 291–330 (2008)
10. Wang, X., Ma, X., Grimson, W.: Unsupervised activity perception in crowded and complicated scenes using hierarchical bayesian models. IEEE Transactions on Pattern Analysis and Machine Intelligence 31(3), 539–555 (2009)

Mixed-Initiative Context Filtering and Group Selection for Improving Ubiquitous Help Systems

Nasim Mahmud, Kris Luyten, and Karin Coninx

Abstract. When people need help they often turn to their social peers for reliable information, recommendation or guidance. It is often difficult to find someone in the vicinity for help or communicate with someone from a distant place who can provide reliable help. Conveying the actual context of the question during remote communication is a cumbersome task, especially when avoiding speech communication. Our approach selects and prioritizes the contextual data for a question, based on the question content. We have developed a prototype for mobile users – the Ubiquitous Help System (UHS) – that implements a mixed-initiative approach for capturing, selecting and prioritizing contextual information as well as for selecting a group of users to send the question. UHS processes the user questions for clues on what context to include and presents its suggestions to the user. Contextual data that can be retrieved using the available sensors on the mobile device is automatically included for sending to the receiving parties alongside the question.

1 Introduction

People often turn to their social peers such as friends, families or colleagues when they need information, any support or guidance to solve a problem. The process of finding and identifying the essential and reliable part of an information or help requires time and effort. It is often convenient to consult a knowledgeable person over searching for the information alone [8]. If we have a specific question from a field in which a friend or colleague is an expert, it is more efficient to consult her or him directly [3, 8].

People are increasingly mobile and therefore, exposed to newer dynamic situations in their everyday lives. That requires them to seek more context dependent

Nasim Mahmud · Kris Luyten · Karin Coninx
Hasselt University - tUL - IBBT,
Expertise Centre for Digital Media,
Wetenschapspark 2, B-3590 Diepenbeek, Belgium
e-mail: {nasim.mahmud, kris.luyten, karin.coninx}@uhasselt.be

P. Novais et al. (Eds.): Ambient Intelligence - Software and Applications, AISC 153, pp. 155–162.
springerlink.com

help, information or guidance in unfamiliar situations. Furthermore, mobile people not only have limited access to their social peers but also have less time to identify as well as justify the information themselves. Most of the cases, unique conditions in real life problems make the Internet an unsuitable source of information. For non–time critical cases, many people utilize their computer mediated online social networks to seek recommendation, help or information as an alternative. People ask their custom query to their social networks such as Facebook [1] by using the status update [9], by asking their question to dedicated social Q&A sites such as Quora [2]. But these networks are not yet suitable for composing questions with rich media and contextual information. Moreover, while on the go, people are less informed about their social network that hinders finding suitable social peers who can provide help and guidance when necessary.

In this paper, we introduce UHS, a social and context aware mobile question and answer system that captures and prioritizes the context information to be dispatched along with the question. It uses audiovisual messages and utilizes the users' context as well as social network to formulate a question. It is unique in a sense that it increases effectiveness by adding context information and increases reliability and comfort by explicitly taking the social network as a source of information and help. Here we also intend to answer the questions: how to classify and prioritize the captured context information to facilitate a user to understand what context information are being dispatched along with the question? How to select a group of potential help providers to send the question to for an optimal outcome? Finally we propose algorithms that addressed these questions.

2 Related Work

Dourish et al. [4] have defined social awareness as understanding activities of others that provides a context for the person's own activity. Such understanding about others allows behavior that is considered natural, socially appropriate, or simply polite. As Fogarty et al. [5] have identified the fact half a decade ago, current computer and communication systems are still largely unaware of the social situations surrounding their usage and the impact that their actions have on these situations. There have been several research attempts, but there is not much development in this regards.

Asking questions and seeking for help from others are between the most common ways for people to solve problems in a social environment. When looking for expert knowledge, people usually seek help from their personal social network [10]. CityFlocks [3] is a newer and context aware mobile system that enables nomadic visitors or new residents in a city to acquire knowledge about the city from the local residents. It also allows the users to share their experiences with the local residents and other users by digitally annotating, commenting and rating any artifacts in the city. It specifically aims to lower the existing access barrier for information. However the system does not consider the contexts of the information seeker and

[1] Facebook, http://www.facebook.com/

[2] Quora, http://www.quora.com/

information provider. VizWiz [2] is one of the most recent question answering system. It allows a blind person to recruit remote sighted workers to help him translate the meaning from an image. The sighted persons recruited by the blind person to answer the question are provided by workers on Amazon Mechanical Turk [3]. It allows composing a question together with a picture of an object of interest and send to the answer providers who are physically apart from the scene. Liu et al. [7] have proposed similar system, also based on crowdsourcing from Amazon Mechanical Turk. Their system, however adds context information such as time and location to compose an image translation job for the Amazon Mechanical Turk. These systems provide no social relation between the person asking the question and the person replying, making the answers an issue of reliability and comfort.

These are few examples of the many studies different 'help systems' that harness help from so called 'human powered service.' However, none focus on the social and context awareness issues to glean help and guidance, which is the focus of our work.

3 Geo-Social Barriers

Distance and Location Barrier. When people are away from their regular social terrain, the geographical distance isolates them from being connected with their social peers. However, by utilizing modern communication technologies such as mobile phones, people can keep themselves connected and partly overcome those barriers. But when someone has a question or needs help, it is difficult for him to find a suitable person in the vicinity or from any distant place who is willing to respond.Fortunately, people can find someone from within their social network by using mobile communication systems. But it requires several attempts before finding the right person who can help. Explicitly communicating with several persons for seeking help increases the value of the so called social factor 'the social cost of seeking help [6]' that people usually try keep at a minimum level.

There are other options to call for help without interrupting the social peers. For example, people often repurpose the use of status message of their social network such as, Facebook. They use it to ask a question or information, seek help or to ask for recommendation [9]. But, it requires substantial amount of time to get a fruitful feedback, which makes this option unsuitable for mobile users. Mobile persons have different requirements than others that lead to information relevance not only being determined by the content but also by extrinsic properties, related to users' current context and urgency. For example, while on the move, they have less time to decide on something.

Interpersonal and Social Barrier. In a natural and social setting people usually turn to their social peers whenever they need some information, help or guidance. When people want to verify or validate their opinion on something, they also seek

[3] Amazon Mechanical Turk, `http://www.mturk.com/` - Last accessed: October 1, 2011 01:30 CET

assistance from their social peers, usually from those with strong-ties. In a collocated social or work environment, people can perceive others' mood or interruptibility relatively easily. And they can ask their social peers for information or seek for help in an acceptable and polite way. But with the increased mobility in modern life, people commute to distant places for work, studies, go for leisure and so forth in a regular basis. This distance plays a vital role for the individuals seeking help from trustable social peers in an effective and comfortable way.

Lack of awareness about social peers increases interpersonal barrier (e.g., weak-ties). However, with modern smart phones, people can share some of their context information to complement the awareness status. For instance, they can share their availability status using instant messenger or social network status updates. But, the question remains how to communicate with necessary context rich information. And what context information is necessary in the given situation. In a face to face communication we can easily exchange a lot of necessary context information without explicitly noticing it.

Determining people's availability status is difficult. It does not only depend on the context that we can measure (e.g., in meeting room), but also depends on some other variables (e.g., who is asking for attention). For instance, one might appear available to reply with the location of a meeting room to a colleague. Whereas at the same time he might remain busy and like to avoid answering a question from a neighbor, even though he has couple of minutes to spare. This results in the fact that an application designer has to address several additional challenges.

When we communicate in person, in a collocated place, we share lot of subtle information that is difficult to capture and process digitally. But, fortunately not all of the context information is as important as others. Abowd and Dey [1] have identified four categories of context information that are important for context aware applications. Those are: location, identity, activity and time. One thing is clear from this category that, the authors have not considered human-human communication, rather they have considered human interaction with the device. In our studies, we are considering the interaction beyond the human's communication with the device to device, rather our discussion is extended up to the level of interaction with another human being. This requires capturing the context of the users from both sides of the communication.

4 Ubiquitous Help System (UHS)

There are more mobile phones than computers. And mobile people need more context aware help and assistance than others. It inspires us to build a social and context aware mobile help system. We have developed the Ubiquitous Help System (UHS), a social and context aware system that helps the user to compose a context rich audiovisual question and to send it to the groups of potential help providers. Additionally, we have developed a desktop Java client for users who wishes to take part in helping others from home or office computer, for example.

The UHS allows a user to take a picture of the subject of interest and allows to ask an audio question. The audio question is converted into text for analyzing its content. The system captures the location, time, and extract urgency factor from the question by analyzing the formerly mentioned text. The system extracts the user's availability status from the to–do list and from the system settings (e.g., ringtone setting) (see figure 1,2,3 and 4) .

For developing the UHS system, the Android platform was chosen so that the system can utilize the hardware sensors such as GPS, camera, voice recorder and so forth that come with most of the Android phones. And it can also utilize several tightly coupled services provided by Google that are accessible through public APIs such as, Google Maps and Google Voice to Text Converter.

5 Our Approach

Context Selection and Prioritization. In order to select the context, the system computes the priority of the context information. It prioritizes the context based on the help seeker's voice input. It accepts a voice question, and then converts the voice into text for being analyzed in the later steps. It finds and matches with all the possible meanings for all the words uttered by the user. For extracting the meaning of the words it utilizes the semantic dictionary WordNet [4]. In short, the WordNet is a social network of words, where words are connected with each other in a 'meaningful' way. Finally after the computation, it suggests the prioritized list of contexts to be dispatched with the question.

The major steps involved are:

1. Tokenize each word of the sentence
2. Find synonym for each token
3. For each synonym find a match
4. If match found, increase the priority of the matched context
5. Sort the list of contexts

Let, V be the spoken audio sentence where S represents the text form of the spoken sentence V. Where, $V \equiv S$
$S = \{ w_1, w_2, \dots w_n | w_i =$ is a word in the sentence $\}$
$w_{jk} = \{ s_{j1}, s_{j2}, \dots s_{jn} | s_{jk} =$ is a synonym of the word $w_j \}$, that is
$S_s = \{ s_{11}, s_{12}, \dots s_{21}, s_{22}, \dots s_{n1}, s_{n2} \dots s_{nn} | s_{jk} =$ is a synonym of a word in $S \}$
$D_l = \{ d_1, d_2, \dots d_n | d_i =$ is an entry in the location dictionary $\}$
$L_c = \{ l_1, l_2, \dots l_n | l_i =$ is an entry in the location context $\}$ and Π is the list of selected context then, $\forall \sigma \in S_s \cup D_l, \quad \exists \lambda \in L_c, \quad then \quad \sigma \equiv \lambda \quad \Rightarrow \quad \lambda \in \Pi$

[4] WordNet, `http://wordnet.princeton.edu/` – Last accessed: January 9, 2012 18:00 CET

Algorithm 1: Context Selection

1: **Input** Text of the speech V.
2: **Input** List of contexts, C.
3: **Output** List of selected context, Π.
4: **while** $|S| > 0$ **do**
5: W $\leftarrow$ Synonym(w_i)
6: D $\leftarrow$ NamedEntity(w_i)
7: $S_s \leftarrow S_s \cup W \cup D$
8: $S \leftarrow S - w_i$
9: **end while**
10: **for** each σ in S_s **do**
11: **for** each c_i in C **do**
12: **if** $\sigma == c_i$ **then**
13: $\Pi \leftarrow$ SelectContext(c_i)
14: **end if**
15: **end for**
16: **end for**
17: **return** Π

Fig. 1 Voice interaction with the device.

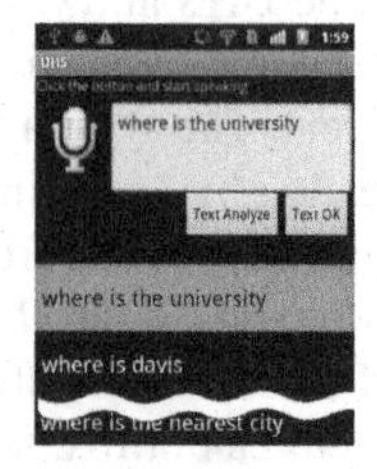

Fig. 3 Voice to text conversion.

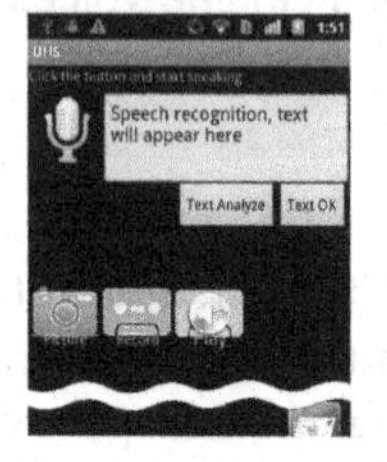

Fig. 2 Main page showing options.

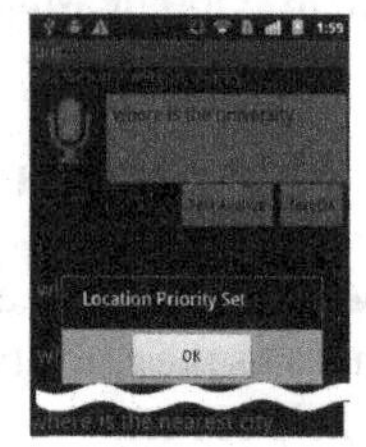

Fig. 4 Context selection.

In other words, if location context L_c, is subset of synonyms of S and the named entity recognition dictionary D_l, then location context is prioritized. In similar fashion, the algorithm analyzes and sets priority to other context information (see algorithm 1).

Group Selection It is difficult for a nomadic user to find a friend within his or her social network who is available and willing to help. To take an example, a person wants reliable help from a friend who is in a distant place. He needs to rely only on ill informed 'guessing.' The person may roughly know that his friend is in a particular office, but he might be busy in a meeting and unable to pick up the phone. The key concept of the group selection algorithm is to select a group of potential help providers from within the social network of the user, who may provide assistance.

UHS takes a mixed initiative approach to select a relevant group of users from the user's social topology. The group selection algorithm takes the selected list of the context, generated by the algorithm described in the previous section. In addition, it takes the task from the shared to-do list of the users. Each task in the to-do list is associated with the location of the task.

Once the system knows the current location of the user over a period of time, it predicts the mode of transport by computing the user's speed. For instance, based on a predefined threshold–a slow movement refers to the user being in a 'walking' state. The idea is that by determining the mode of transport the algorithm can make an assumption of time required to reach the destination. Using this information the 'UsersAffinity' function dynamically binds the user's 'affinity' with the current location (e.g., walking) or with the destination (e.g., driving). From the shared agenda and location of the user's potential friends, it selects persons who are related with the current location or with the destination (see figure 5). In the following step, the algorithm searches for the users who are physically closer and computes who is socially nearby (see algorithm 2). By socially nearby we mean the person with whom the user has frequent interaction.

Algorithm 2: Group Selection

1: **Input** List of selected context, Π.
2: **Input** List of friends from social network, Σ.
3: **Output** Group of selected users, Ψ.
4: **for** each user, $\sigma_i \in \Sigma$ **do**
5: **if** $\sigma_i.status == available$ **then**
6: **if** $\sigma_i.location \in U_c \cup U_d$ **then**
7: /*U_c, U_d: user's current location and destination.*/
8: $\Sigma_s \leftarrow \Sigma_s \cup \sigma_i$
9: **end if**
10: **end if**
11: **end for**
12: $r_l \leftarrow$ UsersAffinity(location, time, U_c, U_d)
13: $\Psi \leftarrow$ RefineList(Σ_s, Π, r_l)
14: **return** Ψ

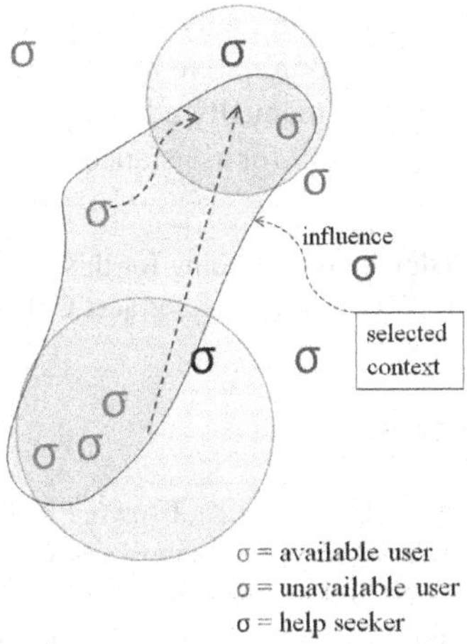

Fig. 5 Group selection. The selection algorithm filtered out some available users who are less relevant to the selected context(Π).

This strategy solves the problem in common cases depending on the availability of required information (e.g., determining availability status of all users). Selecting a suitable group of users completely autonomously is difficult. One of the many reasons behind this is that, in order to determine a suitable group of users, the system should be able to fully understand the user's internal context, which is difficult to measure with available sensors. However, when the automatically selected group does not reflect the user's intention, he can override the selection by manually editing the list of users to broadcast the question. This semiautomatic approach is beneficial because the user can see who is relevant within the current context and current query. He can intervene the selection, if necessary. This flexibility is required to obtain an intelligent and comfortable solution.

6 Conclusion

In this paper we have demonstrated how the Ubiquitous Help System (UHS) can be used to assist a help seeker to frame a query for assistance with the appropriate context and within the seekers social context. We addressed two research questions here first, how to classify and prioritize the captured context information to facilitate a user to understand what context information is being dispatched along with the question; and second how to select a group of potential help providers to send the question to for an optimal outcome. We have presented an algorithm that identifies relevant context based on the question and an algorithm that groups reliable social peers to assist a help seeker. Our algorithms select the appropriate groups of users based on relevant context information for a question. The context information is automatically extracted from sensors that are available on typical Android phones (e.g. GPS) and from analyzing the content of the voice question of the user. Furthermore, a mixed initiative user interface provides the necessary means to the user to intervene and manually add, remove or edit existing context information before distributing a query for assistance.

Acknowledgements. Funding for this research was provided by the Research Foundation – Flanders (F.W.O. Vlaanderen, project CoLaSUE, number G.0439.08N).

References

1. Abowd, G.D., Dey, A.K., Brown, P.J., Davies, N., Smith, M., Steggles, P.: Towards a Better Understanding of Context and Context-Awareness. In: Gellersen, H.-W. (ed.) HUC 1999. LNCS, vol. 1707, pp. 304–307. Springer, Heidelberg (1999)
2. Bigham, J.P., Jayant, C., Ji, H., Little, G., Miller, A., Miller, R.C., Tatarowicz, A., White, B., White, S., Yeh, T.: Vizwiz: nearly real-time answers to visual questions. In: Proceedings of the 2010 International Cross Disciplinary Conference on Web Accessibility (W4A) (2010)
3. Bilandzic, M., Foth, M., De Luca, A.: Cityflocks: designing social navigation for urban mobile information systems. In: Proceedings of the 7th ACM Conference on Designing Interactive Systems (2008)
4. Dourish, P., Bellotti, V.: Awareness and coordination in shared workspaces. In: Proceedings of the 1992 ACM Conference on Computer-Supported Cooperative Work (1992)
5. Fogarty, J., Hudson, S.E., Atkeson, C.G., Avrahami, D., Forlizzi, J., Kiesler, S., Lee, J.C., Yang, J.: Predicting human interruptibility with sensors. ACM Trans. Comput.-Hum. Interact. 12, 119–146 (2005)
6. Lee, F.: The social costs of seeking help. The Journal of Applied Behavioral Science 38(1) (2002)
7. Liu, Y., Lehdonvirta, V., Kleppe, M., Alexandrova, T., Kimura, H., Nakajima, T.: A crowdsourcing based mobile image translation and knowledge sharing service. In: Proceedings of the 9th International Conference on Mobile and Ubiquitous Multimedia (2010)
8. McDonald, D.W., Ackerman, M.S.: Expertise recommender: a flexible recommendation system and architecture. In: Proceedings of the 2000 ACM Conference on Computer Supported Cooperative Work, pp. 231–240. ACM, New York (2000)
9. Morris, M.R., Teevan, J., Panovich, K.: What do people ask their social networks, and why?: a survey study of status message q&a behavior. In: Proceedings of the 28th Int. Conf. on Human Factors in Computing Systems (2010)
10. Nardi, B.A., Whittaker, S., Schwarz, H.: It's not what you know, it's who you know: Work in the information age. First Monday 5 (2000)

Designing a Middleware-Based Framework to Support Multiparadigm Communications in Ubiquitous Systems

Carlos Rodríguez-Domínguez, Tomás Ruiz-López,
Kawtar Benghazi, and José Luis Garrido

Abstract. Ubiquitous systems require to support key functionalities such as context-awareness, cooperation and proactivity, making use of different communication paradigms (Request/Response, Pub/Sub, etc.). Each paradigm is usually supported by specific architectures (such as agent-based, event-driven, P2P and SOA), communication mechanisms (RPC, broadcasting, etc.), and technologies (middlewares, MANETs, programming languages, etc.). This paper presents a conceptual design of a technology-independent, middleware-based framework intended to reduce the inherent complexity of developing software solutions in an integrated way. The framework is capable of integrating heterogeneous communication-related components.

Keywords: Ubiquitous systems, middleware, software framework, context-aware applications, reusability, ease of development.

1 Introduction

The commercial success of mobile devices has initiated the transition from traditional computing environments, where devices appear situated in a very specific place and whose user interaction is "intrusive", to ubiquitous computing environments, where devices are available everywhere and whose interaction should be "as much natural as possible" [13].

Also, in ubiquitous computing, software applications are increasingly supporting *context-awareness*, that is, the ability to adapt their functionalities according to the information that characterizes the situation of a person, place or object involved in the interaction between the user and the application itself [3] (physical location, nearby resources, the user identity, etc.). Supporting context-awareness usually

Carlos Rodríguez-Domínguez · Tomás Ruiz-López · Kawtar Benghazi · José Luis Garrido
Department of Computer Languages and Systems, University of Granada,
C/ Periodista Daniel Saucedo Aranda S/N, 18014 Granada, Spain
e-mail: {carlosrodriguez,tomruiz,benghazi,jgarrido}@ugr.es

P. Novais et al. (Eds.): Ambient Intelligence - Software and Applications, AISC 153, pp. 163–170.
springerlink.com

involves the interoperation between non-collocated entities and the design or adoption of concrete communication schemes and technologies. Both tasks are considered complex, since it is usually required to make a continuous, location-independent use of context-aware applications and to exchange information with other users (or services) irrespectively of the technologies that they use.

In this context, middleware-based solutions are mainly accepted to overcome interoperability issues, also providing portability between different underlying platforms and sometimes satisfying other quality properties (efficiency, scalability, security, and so forth). Commonly, in ubiquitous computing, and more particularly, in Ambient Assisted Living (AAL) systems, it is necessary to satisfy several requirements on the basis of adopted design decisions and selected technologies, hence requiring the integration of several middleware technologies [7]. In practice, this integration increases the development complexity since it is required to deal with different, non-integrated technologies. Thus, we believe that an integration of distinct middleware technologies under a conceptual, unifying and highly-flexible approach would decrease the difficulty of dealing with their technological differences.

Furthermore, in order to facilitate the development of applications, software frameworks usually turn out to be a common choice in software engineering [4]. Software frameworks comprise a set of hot and frozen spots. Hot or flexible spots represent the abstractions that enable to adapt the functionalities of the framework to the specific requirements of a system, whereas frozen or stable spots define basic components that remain unchanged in any instantiation of the framework [9]. A framework also provides "inversion of control" mechanisms, that is, the framework is responsible of executing the instantiations of the hot spots when required, therefore improving software reusability by decoupling the mechanisms to control the software execution from the specific implementations of the hot spots.

This research work devises the conceptual design of a technology-independent, middleware-based framework identifying abstractions to be associated with the mechanisms available in a framework (hot spots, frozen spots and inversion of control). The aim is to decrease the complexity of developing ubiquitous systems by providing more technology-independent abstractions that can be instantiated "on-demand" with specific communication technologies. This work is particularly applicable to AAL systems, which usually require to communicate different entities (services, applications, agents, etc.) with different paradigms (PubSub, RPC, P2P, etc.) and through different communication technologies (BlueTooth, Wi-Fi, etc.) so as to provide context-awareness, proactivity and cooperation support.

The remainder of the paper is structured as follows. In Section 2, some work that is related to the proposal described in this paper is briefly analyzed. Section 3, presents the proposed framework. Finally, Section 4 exposes some concluding remarks.

2 Related Work

A small number of research works have previously addressed the problem of decreasing the complexity of developing integrated ubiquitous systems based on heterogeneous communication schemes, technologies, architectures and paradigms.

Getov et al. [6] propose a multiparadigm solution to deal with communications in grid computing. This proposal deals with message exchanging, event distribution and dynamic discovery of shared resources. In addition to be oriented to grid computing, the main difference between this approach and the proposal presented herein is that it is specifically oriented to Java programming, whereas our proposal is independent of a language and a technology, thereby separating the implementation of the entities (applications, services, agents, and so on) from specific communication technologies.

UIC-CORBA [11] is a reflective middleware supporting different communication paradigms and mobile devices. Its architecture is based on CORBA, making it complex to instantiate it on the basis of more modern middleware solutions for ubiquitous systems, like Mobile-Gaia [12], which supports dynamic discovery of nearby resources through BlueTooth and IrDA discovery protocols.

OpenORB [2] is a middleware to deal with message-oriented and event-based communications in distributed systems. It supports the configuration of several types of middleware instances due to its reflective capabilities. However, this solution is not oriented to ubiquitous systems and it does not cope with context management or dynamic discovery of resources.

SeDiM [5] is a middleware framework to allow heterogeneous service discovery protocols to interoperate. The work presented herein complements this work by also incorporating models to support message exchanging, event distribution and context management between heterogeneous applications, services, etc. Also, in contrast with other middleware solutions specifically developed for ubiquitous systems, our proposal aims to be more technology-independent and to offer a higher abstraction level over the supported functionalities (event distribution, discovery of nearby devices, and so on)

3 A Framework Design Oriented to Multiparadigm Communications in Ubiquitous Systems

The extra difficulty of incorporating multiple communication paradigms in ubiquitous systems should be balanced with a better availability of powerful techniques providing the needed abstractions to manage development complexity. This research work proposes a middleware-based framework on the basis of several technology independent models related to the basic communication functionalities that should be present in ubiquitous systems [8]: (1) Message exchanging, that is, one-to-one communications between context-aware and services; (2) Distribution of events to notify state changes in an application to other applications; (3) Dynamic discovery of nearby entities to exchange messages or notify events; (4) Retrieval of contextual information about the physical environment of a user. For example, these functionalities are basic to support proactivity and cooperation in AAL systems.

Fig. 1 outlines the logical architecture of the proposed framework, consisting of a set of mechanisms related to the communication functionalities identified above in order to promote an easier and more complete development of software solutions for ubiquitous systems.

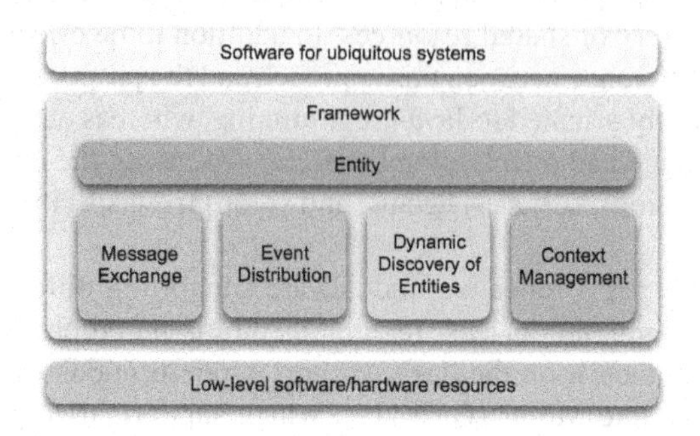

Fig. 1 Overview of the logical architecture of the proposed framework

An entity in this logical architecture, independently of its nature (a service, an application, an agent, etc.) and the selected development technologies, should provide basic communication functionalities associated with message exchange (chats, videoconference, etc.), event distribution (coordination in mobile shared workspaces, etc.), dynamic discovery of entities (spontaneous social networks, location-aware ad systems, etc.) and context management (sensor networks in AAL systems, etc.). The following subsections describe the technology-independent models in the framework to deal with these communication functionalities.

3.1 Message Exchanging

Message exchanging is commonly used for one-to-one communications between applications and services. The wide range of requirements associated with ubiquitous systems usually involves integrating several existing technologies and protocols, which usually involves higher development complexity. The technology independent model depicted in Fig. 2 is intended to overcome this drawback. The elements represented in this model are described in Table 1.

Message and *Marshallable* hot spots can be instantiated with different communication protocols to exchange information between heterogeneous *Proxies* and *Servants* through one or more *CommunicationDevices*. These technology independent concepts increase the flexibility of the proposed framework enabling the integration of different technologies, while promoting the separation between the operational logic and the specific underlying technologies.

In order to be able to integrate communication technologies as needed, *Device Pool* frozen spot is provided. This pool stores *CommunicationDevice* instances and

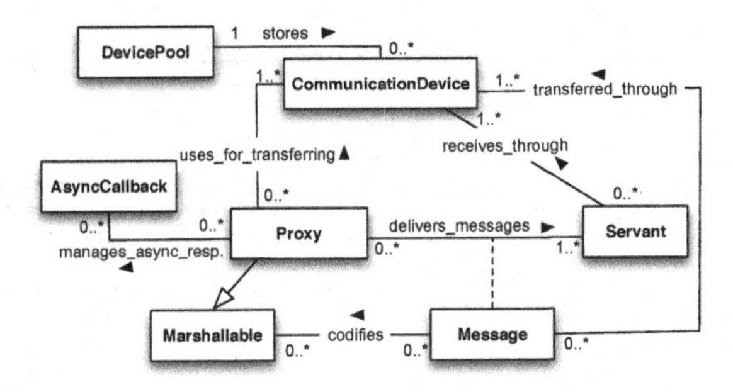

Fig. 2 Message exchanging model

Table 1 Frozen and hot spots related to message exchanging in the proposed framework

Hot Spot	Description
Message	Information exchanged between a sender and a receiver (usually between applications and services).
Communication Device	Abstract definition of a communication technology (for instance, BlueTooth or Wi-Fi) intended to exchange messages.
Servant	Model of a remotely accesible object.
Proxy	Model of a binding to a servant, to "transparently" access remote objects.
Marshallable	Model of a codifiable object that can be transferred as part of a message.
Async. Callback	Action to be executed whenever an asynchronous message is received.
Frozen Spot	**Description**
Device Pool	Stores instantiated *communication devices* and provides one in case others are not available

automatically provides one at run-time, in case others are not available, so as to provide mobility support. Furthermore, it incorporates several mechanisms to support both "off-line" and "on-line" operation modes within. The goal of these mechanisms, in general, is to transparently store messages that can not be delivered and to try to send them whenever it is possible.

3.2 Event Distribution

Entities may communicate changes in their internal state to a set of other interested entities, thus requiring event distribution. For example, in a domotic environment, if a light state changes from *off* to *on*, then an event should notify applications of this occurrence in order to reflect this new state in their corresponding UI's. In Fig. 3, the devised event distribution model, based on PubSub paradigm, is depicted using a UML class diagram.

Previous models aim to support event distribution independently of the technology that is used to implement it. For example, this model could be instantiated using

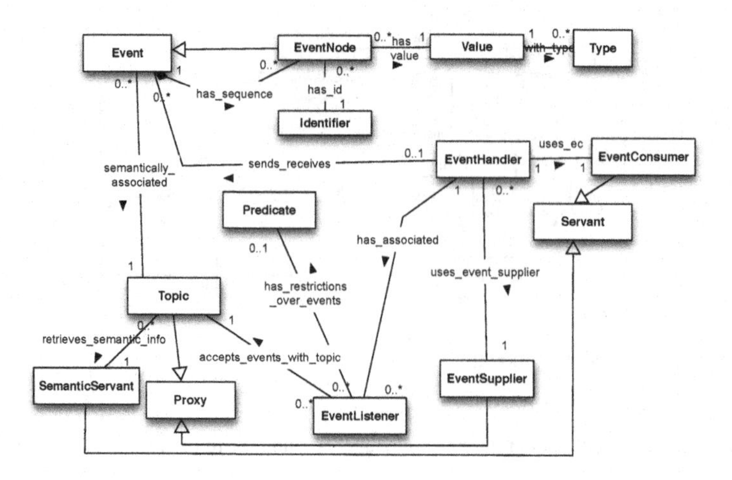

Fig. 3 Event distribution model

CORBA Notification Service or DDS. Technology independence is mainly achieved by instantiating *EventConsumer* and *EventProducer* hot spots. *Events* are modeled as a collection of *event nodes*, each of them associated with an identifier and a typed value. Events are associated with semantic information (*topics*) (like in DDS specification), which is stored in a *Semantic Servant*.

The elements that provide support for publishing and subscribing to event topics are *EventListeners* and the *EventHandler* (a frozen spot). An *EventListener* allows establishing the actions to be executed whenever a notified event is related to a specific topic and accomplishes a specific *Predicate*. The *EventHandler* is responsible for receiving published events and delivering them to the appropriate listeners. The proposed *Event* model is detailed in [10].

3.3 Dynamic Discovery of Entities

In ubiquitous systems, entities should be able to dynamically discover other entities that are physically placed around them (or that are accessible through the underlying communication technology) in order to exchange information. *How* entities are discovered depends specifically on the underlying technology. A model is proposed to support dynamic discovery of entities using different underlying technologies. This model consists of a *mobile discoverer*, which specifies a method for dynamically discovering entities, and a *mobile discovery listener*, which asynchronously triggers actions whenever a *mobile discoverer* provides new results.

The separation between the implementation of the specific discovery algorithm and the actions to be executed whenever an entity is discovered or removed from a specific environment allows improving maintainability (i.e., discovery algorithm should not be related to the specific discovery-related actions to be executed) and reusability (i.e., actions are usually independent of a specific communication technology and could be reused between discovery algorithms). In addition, we consider that the actions should be always executed asynchronously, since it is not possible to limit the time needed for executing most of the existing discovery algorithms and the discoverer entity should never stop its execution flow.

3.4 Context Management

Context management is one of the key functionalities that a framework for developing context-aware applications and services for ubiquitous systems should provide. According to Dey et al. [3], the context is the information that characterizes the situation of a person, place or object involved in the interaction between a user and an application. In this work, the information that we consider more relevant to characterize the context is: (1) the contextual information provided by the available services in a specific location; (2) the entities (services or applications) that are physically around each specific user; (3) the topics associated with each notified event, as it indirectly provides information about the tasks that each entity is carrying out.

By taking into account the previously mentioned parameters, the framework provides a *Context Manager* hot spot, so as to enable the possibility to include different application-specific aspects of context management. For example, user behaviour or specific climatic conditions could be considered crucial information in some applications. In the proposed framework, the *Context Manager* is a *proxy* (see Section 3.1) that can have bindings to several services providing contextual information (location, weather forecast, etc.). It also incorporates a *mobile discoverer* hot spot (see Section 3.3) to discover nearby users and includes a binding to the *semantic servant* (see Section 3.2) to retrieve semantic information about notified events, which offers indirect information about other users' tasks.

4 Concluding Remarks

The software framework design described in this paper provides abstractions to support an easier integration of several communication paradigms commonly used in ubiquitous systems, independently of the specific underlying communication technologies. This is particularly relevant to AAL systems, which usually require to integrate different technologies in order to support different required functionalities, such as context-awareness, proactivity and cooperation, with different architectures (agent-based, event-driven, P2P, etc.) and communication paradigms (PubSub, RPC, etc.). Moreover, the provided conceptual support allows decreasing the development complexity that entails the incorporation of heterogeneous technologies, architectures and paradigms. The framework provides a set of hot and frozen spots related to four communication functionalities: message exchanging, event distribution, dynamic discovery of nearby entities and context management. These hot spots can be instantiated on top of existing middleware solutions, and are highly reusable. The benefits of the framework also contribute to improve the maintainability, since the usage of specific technologies is isolated from the actual implementation of the software solutions, making it easier to detect or mitigate development deficiencies and to meet new requirements in the future. Finally, it is important to note that an opensource middleware called BlueRose [1] has been implemented on the basis of the framework to demonstrate its feasibility to develop highly reusable software. It has been used for the development of several context-aware applications and services.

Acknowledgements. The Spanish Ministry of Education funds this research work through projects TIN2008-05995/TSI and TIC-6600.

References

1. BlueRose, `http://code.google.com/p/thebluerose`
2. Coulson, G., Blair, G., Parlavantzas, N., Yeung, W.K., Cai, W.:: A Reflective Middleware Approach to the Provision of Grid Middleware. Middleware for Grid Comp. (2003)
3. Dey, A.K.: Understanding and Using Context. Journal of Personal and Ubiquitous Computing 5(1), 4–7 (2001)
4. Fernández-Riverola, F., González-Peña, D., López-Fernández, H., Reboiro-Jato, M., Méndez, J.R.: A JAVA application framework for scientific software development. Software: Practice and Experience (2011)
5. Flores, C., Grace, P., Blair, G.S.: SeDiM: A Middleware Framework for Interoperable Service Discovery in Heterogeneous Nets. ACM Trans. Auton. Adapt. Syst. 6(1) (2011)
6. Getov, V., von Laszewski, G., Philippsen, M., Foster, I.: Multiparadigm Communications in Java for Grid Computing. Communications of the ACM 44(10), 118–125 (2001)
7. Grace, P., Blair, G.S., Samuel, S.: A reflective framework for discovery and interaction in heterogeneous mobile environments. ACM SIG. Mob. Comp. and Comm. 9(1) (2005)
8. Maia, M., Rocha, L., Andrade, R.: Requirements and challenges for building service-oriented pervasive middleware. In: Procs. of the 2009 Intl. Conf. on Pervasive Services (2009)
9. Pree, W.:: Design Patterns for OO Software Development. Addison-Wesley (1995)
10. Rodríguez-Domínguez, C., Benghazi, K., Noguera, M., Bermúdez-Edo, M., Garrido, J.L.: Dynamic Ontology-Based Redefinition of Events Intended to Support the Communication of Complex Information in Ubiquitous Computing. Journal of Network Protocols and Algorithms 2(3), 85–99 (2010)
11. Román, M., Kon, F., Campbell, R.H.: Reflective Middleware: From Your Desk to Your Hand. IEEE Distributed Systems Online 2(5) (2001)
12. Shankar, C., Al-Muhtadi, J., Campbell, R., Mickunas, M.D.: Mobile Gaia: A Middleware for Ad-Hoc Pervasive Computing. In: IEEE Consumer Comm. and Networking Conf. (2005)
13. Weiser, M.: The computer for the 21st century. Scientific American 265(3), 94–104 (1991)

A Two-Stage Corrective Markov Model for Activities of Daily Living Detection

Love Kalra, Xinghui Zhao, Axel J. Soto, and Evangelos Milios

Abstract. In this paper we propose a two-stage, supervised statistical model for detecting the activities of daily living (ADL) from sensor data streams. In the first stage each activity is modeled separately by a Markov model where sensors correspond to states. By modeling each sensor as a state we capture the absolute and relational temporal features of the atomic activities. A novel data segmentation approach is proposed for accurate inferencing at the first stage. To boost the accuracy, a second stage consisting of a Hidden Markov Model is added that serves two purposes. Firstly, it acts as a corrective stage, as it learns the probability of each activity being incorrectly inferred by the first stage, so that they can be corrected at the second stage. Secondly, it introduces inter-activity transition information to capture possible time-dependent relationships between two contiguous activities. We applied our method to three ADL datasets to show its suitability to this domain.

Keywords: Activities of Daily Living, Markov Model, HMM.

1 Introduction

Automatic detection of daily human activities can greatly benefit the in-home elder care and home automation systems, allowing the elderly to live independently in the privacy of their own homes. In assisted-living complexes, context-aware systems can help monitor the status of the elderly occupants. Human activity detection and recognition are not trivial tasks. There are several technical challenges and ethical responsibilities to perform the monitoring effectively without compromising a person's privacy. Thereby, in this work we focus on non-visual sensors to detect the activities.

Love Kalra · Xinghui Zhao · Axel J. Soto · Evangelos Milios
Faculty of Computer Science, Dalhousie University
6050 University Avenue, Halifax, NS, Canada B3H4R2

P. Novais et al. (Eds.): Ambient Intelligence - Software and Applications, AISC 153, pp. 171–179.
springerlink.com

Given the contextual and temporal nature of human activities in a typical home setting, various approaches have been proposed for detecting the activities using non-visual sensors. A context-aware Hidden Markov Model (HMM) based adaptive ADL system proposed in [1] is capable of performing online learning and updating the model providing higher precision than conventional HMM when the distance between the training and testing set is large. Another interesting approach is proposed in [7] where the user-object interactions are exploited to infer the activities. In [3, 7] RFID tagged objects and other sensors are used to capture the activities and evaluating different aspects of activity detection, such as the effectiveness of each type of sensor to detect an activity independently. In [6] a traditional layered HMM model is used to infer the human actions on a desk in an office environment at different time granularities. A novel two phase approach is presented in [12]where the focus is on detecting the abnormal human activities to provide support to the occupant of a house or other premises in case of emergency. In video-based activity recognition approaches, human body postures and locations are used to infer the activities. Mori et al. used hierarchical structure of actions to capture the fine details of human postures using continuous HMM [4].

HMMs, where all the activities are modeled as states, has been the most popular approach [1, 11] to model ADLs. Although modeling the activities with HMM in this manner has been shown to be feasible, it lacks of the flexibility to make use of the temporal features at different time scales. Such a HMM fails to capture the temporal information within the activities, also known as intra-activity information. The intra-activity information is very important for distinguishing the activities that trigger the same set of sensors but have different triggering patterns.

To capture this low-level information, we propose a two-stage model. In the first stage each activity is modeled as a Markov chain. Although this configuration helps the model to capture the intra-activity information, it creates some other challenges. First, when modeling the activities independently, we lose the inter-activity transitions. Second, we require a method to segment the data by recognising the points in time where an activity ends and a new one begins. The accurate segmentation of the data is a challenging problem. When modeling each activity individually, the effective decoding of the activities requires the data segments to consist of instances from only one activity. There is a rather simple approach taken by [2] where sensor data streams are segmented using fixed-length window. This approach does not consider the start and end times of the activities and may generate segments with more than one activity. Tapia et al. [9], on the other hand, use a sliding window technique with a window whose length is equal to the average length of the targeting activity. Although this method provides better segmentation, it may be problematic for detecting instances of activities with a length different from the mean and it also requires repeated scans of the data stream for each activity. In Section 2.1 we propose a novel approach for the segmentation problem which addresses these challenges.

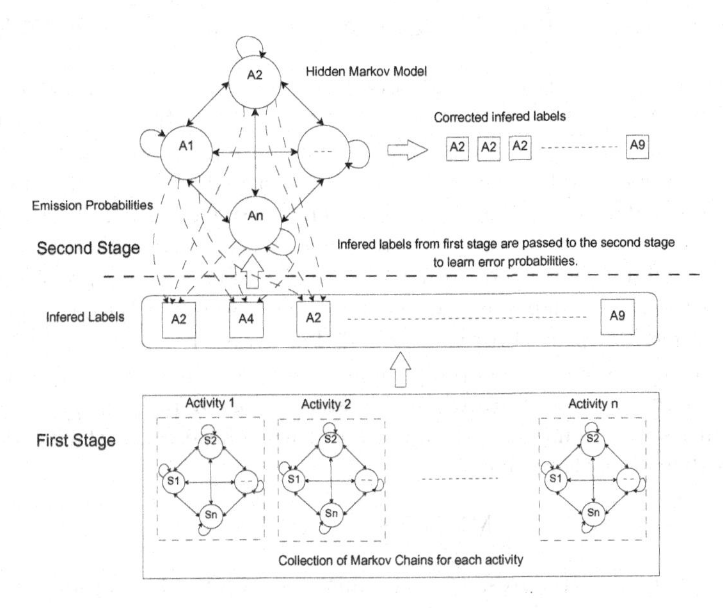

Fig. 1 Model architecture. The first stage is a collection of activity Markov chains and the second stage is a corrective HMM that gives the final detected activities.

The inter-activity information that is not modeled in the first stage is captured at the second stage. The second stage consists of a HMM where each activity is modeled as a state, but the observation distribution is based on the first-stage output (Fig. 1). The second stage aims at correcting possible misclassification originated in the first stage. This is done by learning the misclassification pattern from the first stage. This stage, in addition considers the relative temporal relationship between activities, which tends to improve the confidence of the model when inferring the activities.

2 Multi-Markov Model (MMM)

To model the activities more accurately and make better use of the information available in the form of relative and absolute temporal features, we model the activities at a finer time scale in the first stage. The temporal features between the activities are introduced in the second stage. The architecture of the proposed two-stage model is shown in Fig. 1. Markov chains are suitable for modeling time-varying discrete stochastic processes [5]. In addition, they support the inclusion of temporal and spatial characteristics of the activities discussed above, either implicitly or explicitly. In this work we have adopted the Markov chain framework as described in [2].

2.1 Multi-Markov Model - First Stage (MMM-FS)

The first stage consists of an individual Markov chain for each activity. To detect an activity, we feed a segment of sensor data stream corresponding to that activity to each activity model. The activity model that gives the highest likelihood value is selected as the most probable activity. In this stage we consider different activities to be independent of each other. Each activity Markov chain consists of a set of states, $S = s_1, s_2, ..., s_n$, where each state is a sensor that is activated during the activity. The activity process starts in one of the states, depending upon the initial state distribution, and steps from one active sensor state to another. The transition between the states s_i to s_j at any given time step depends on the probability a_{ij} which in turn depends only on the last state of the system (Markov property). Transition probabilities a_{ij} are modeled during the learning phase. The activity Markov chain is defined by three parameters:

$$\text{MMM-FS} = (S, \Pi, A) \tag{1}$$

where S is the set of states, Π is the initial state distribution and A is the state transition probability matrix.

During the learning phase, each activity is modeled by learning the distributions Π and A from the training samples using the following equations:

$$a_{ij} = T_{ij}/T_i \tag{2}$$

$$\pi_i = N_i/N \tag{3}$$

where T_{ij} is the number of i to j state transitions, T_i is the number of transitions from state i, N_i is the number of activations of state i and N is the number of states. These transitions capture the sequences of sensor activations.

In the testing or decoding phase, a test sequence is fed to each of the activity model and the maximum likelihood of each model given the observation is evaluated using the Viterbi algorithm [8]. The Viterbi algorithm reduces the problem of finding the most likely state sequence given a model into finding the shortest path from the start state to the end state in a trellis of all the possible paths given the model. The activity model with the highest maximum likelihood value is chosen as the activity for the given test sequence of the sensor data.

The Maximum-Likelihood rate of change (MLROC) segmentation technique proposed in this work is inspired by the observation that the likelihood value of the most likely activity changes substantially when the transition from one activity to another occurs. This is due to the fact that different sensor triggering patterns occur when a new activity starts. Based on this observation we increment the length of the window one unit time at each step while monitoring the rate of change in the likelihood value using

Eq. 4, where $\hat{\theta}$ is the maximum likelihood value. When the rate of change value exceeds a predefined threshold value, the window is terminated and a new open window is initiated from the next time step.

$$\textit{Rate of Change} = (\hat{\theta}(t) - \hat{\theta}(t-1))/\hat{\theta}(t-1) \tag{4}$$

To select the rate of change threshold, we measure the alignment accuracy of the segmented data with the original labels during the training phase. The alignment is measured by calculating the distance from the start and end point of the identified segments with the start and end points of the original labels. To quantify the best segmentation we designed an ad-hoc distance measure that was used to evaluate the best threshold value. This evaluation is done over a validation set and is then used for the testing set.

2.2 Multi-Markov Model - Second Stage (MMM-SS)

The first stage does not model the relationships between the activities executed over the day because each activity is modeled individually. Therefore, we add a second stage where the relative temporal features, at the activity level, are modeled. In addition, the second stage boosts the accuracy of ADL recognition by acting as an error-correction layer on top of the first stage. We chose HMM for the second stage, where the activity process that is corrected by the second stage is modeled as hidden variable and the activities that are infered from the first stage act as observations for the second stage. The error-correction process using HMM works as follows:

In the training phase, the corrected activity process is modeled by learning the activity transitions from the training dataset and the error probabilities from the first stage output. In the testing phase, the Viterbi algorithm is used to find the corrected activity sequence using the transition for each activity and the error probabilities from the first stage. To illustrate the error correction process of the second stage consider the output activity sequence of the first stage as $A2, A4, A2, \ldots\ldots, A9$, shown in Fig. 1. Here, the second activity in the sequence is incorrectly predicted as $A4$ which originally was an activity $A2$. During the training phase of the second stage the model learns the probability of first stage to classify activity $A2$ as activity $A4$ and then use this information along with the transition parameter to correct the activities at the second stage. An error-correction HMM is defined by a five-tuple:

$$\text{MMM-SS} = (S, O, \Pi, A, E) \tag{5}$$

where S is the set of hidden states (activities), O is the set of observation symbols (activities learned from the first stage), Π is the initial state distribution, A is the state transition probability distribution matrix and E is a matrix where i, j element is the probability of i-th activity being classified as activity j.

3 Datasets and Experiments

3.1 Datasets

In this work we used two different datasets. Both datasets are collected from two different houses with single occupancy. The occupant of each house is asked to perform a certain set of activities during the day. For modeling and evaluation purposes both datasets provide activity labels.

The first dataset (DS-1) [10], consists of 18-day data of 21 binary sensors installed in a two storey house. The sensors are installed in everyday objects in the house, such as reed switches in the cabinets and doors, mercury contact for the movement of objects such as drawers and a float switch for the toilet flush system. The house is occupied by a 57 year old male, who was asked to perform fifteen fixed every day activities. The activities were hand annotated by the subject in a diary.

The second dataset (DS-2) [9] uses the Mites wireless sensor nodes designed by the MIT Placelab. These sensor nodes can be equipped with different sensor types. This dataset, however, consists of only reed switches that are connected to data collection boards. The sensors are installed in everyday objects such as drawers, refrigerator and containers. This dataset consist of the data from two different houses occupied by two different subjects.

3.2 Experiments

We performed two sets of experiments. The first set of experiments demonstrates the performance of the proposed MLROC segmentation technique, while the second set of experiments shows the performance of our two-stage approach for activity detection.

To evaluate the performance of the proposed segmentation method, we measure the alignment accuracy of the resultant data sequence segments. The alignment accuracy is calculated using a distance measure. This distance measure calculates the distance of misalignment between the end points of the segmented and the original labels.

Due to the small size of the datasets, we tested our two-stage approach for activity detection on each day in a leave-one-day-out fashion. We evaluated the activities by two measures: average time slice accuracy and average class accuracy. The average time slice accuracy is the proportion of time instances where an activity is classified correctly and then averaged over all the days in the dataset. The average class accuracy is the proportion of instances that were correctly classified during the activity and then averaged over all the days for that activity. We present the average time slice and average class accuracy results for our approach in comparison with the approach presented in [9] and with regular single stage HMM.

Table 1 Activity identification accuracy for DS-1.

Accuracies	MMM-FS ROC = 90(%)	MMM-SS ROC = 90(%)	HMM(%)
Class Accuracy	58.33	61.22	55.5
Time Slice Accuracy	89.06	89.38	87.03

4 Results

To compare our approach with the fixed window technique that is commonly used for the data segmentation, we find the average distance over 16 days on DS-2 (subject 1) dataset. For fixed window lengths of 2, 7, 10 and 15 minutes the average distances were 1590.4, 520.13, 390 and 256.1. The decreasing distances indicates that the majority of the activities over the days spans through longer periods. For MLROC we test the threshold values of 80, 90, 95 and 99 and we got the distances of 150, 103, 92.2 and 112. Based on these distances we chose 95 to be the threshold value for DS-2 (subject 1) dataset.

Tables 1, 2(a) and 2(b) show the comparative results from our two-stage approach MMM-FS (first stage) and MMM-SS (second stage) with HMM and Multi-Naive Bayes classifier (MNBC) as discussed in [9]. It can be observed that for almost all the activities our model performs better than the MNBC. When comparing to HMM (Tables 2(a) and 2(b)) some activities

Table 2 Class and Time-Slice accuracy for DS-2.

(a) Subject-1.

Activities	MNBC(%)	MMM-FS(%) ROC = 95(%)	MMM-SS(%) ROC = 95(%)	HMM(%)
Prep. Lunch	29	33.03	42.67	53.76
Toileting	31	44.32	42.69	72.3
Prep. Breakfast	6	37.05	23.58	48.24
Bathing	29	43.02	50.47	27.36
Dressing	3	65.43	66.79	27.67
Grooming	26	46.47	54.29	31.5
Prep. Beverage	13	21.28	23.84	16.84
Doing Laundry	7	63.37	63.15	21.05
Time Slice Accuracy	-	41.82	45.75	43.68

(b) Subject-2

Activities	MNBC (%)	MMM-FS ROC = 99(%)	MMM-SS ROC = 99(%)	HMM (%)
Prep. Dinner	30	37.42	44.1	6.23
Prep. Lunch	22	34.32	29.08	16.62
Listening to music	9	54.3	55.67	28.85
Toileting	23	32.13	36.84	70.45
Prep. Breakfast	24	13.56	16.93	56.71
Washing Dishes	11	34.36	31.32	29.92
Watching TV	16	43.39	43.15	23.22
Time Slice accuracy	-	37.17	37.12	33.77

are better identified with our approach while others with HMM. Due to the presence of a highly variable distribution of activities between the different days it was not possible to prove whether the difference between MMM and HMM was statistically significant. Effects of this variable nature of the activities are observable in cases where the accuracy degrades from first stage to second stage. However, certain activities, namely *grooming*, *dressing* and *doing laundry* in subject 1 and *preparing dinner*, *Watching TV* in subject 2 showed significantly better accuracy with MMM-SS than with HMM with a probability error lower than 0.05.

5 Conclusions

ADL detection is a very challenging problem. The high variability in the nature of activities, sensor noise and annotation inaccuracies, all contribute to the complexity of modeling the ADL effectively. In this work, we proposed a two-stage approach to decompose the problem into two levels. The first stage models the activities at their atomic level and the second stage attempts to correct the misclassified information from the first stage by learning the inter-activity transitions from the data and the error distribution from the first stage output.

We have demonstrated that our approach gives a higher activity identification accuracy than MNBC in almost all test cases. The experiments also suggest that, although decomposing the activities help to improve the detection accuracies in most cases, it can also degrade detection accuracy for certain activities. We have also shown that traditional HMMs are surprisingly competitive under such complex distribution and presence of noise. Finally, we proposed a novel segmentation approach that provides better segmentation of the data and hence allows better classification.

Acknowledgments. The financial support of Kanayo Software Inc. and MITACS is gratefully acknowledged.

References

1. Cheng, B.-C., Tsai, Y.-A., Liao, G.-T., Byeon, E.-S.: HMM machine learning and inference for activities of daily living recognition. The Journal of Supercomputing 54(1), 29–42 (2009)
2. Hasan, M.K., Rubaiyeat, H.A., Yong-Koo, L., Sungyoung, L.: A reconfigurable HMM for activity recognition. In: 10th International Conference on Advanced Communication Technology, pp. 843–846 (2008)
3. Logan, B., Healey, J., Philipose, M., Tapia, E.M., Intille, S.S.: A Long-Term Evaluation of Sensing Modalities for Activity Recognition. In: Krumm, J., Abowd, G.D., Seneviratne, A., Strang, T. (eds.) UbiComp 2007. LNCS, vol. 4717, pp. 483–500. Springer, Heidelberg (2007)

4. Mori, T., Segawa, Y., Shimosaka, M., Sato, T.: Hierarchical recognition of daily human actions based on continuous hidden markov models. In: 6th IEEE International Conference on Automatic Face and Gesture Recognition, pp. 779–784 (2004)
5. Norris, J.R.: Markov Chains. Cambridge University Press (1998)
6. Perdikis, S., Dimitrios, T., Strintzis, M.G.: Recognition of humans activities using layered hidden Markov models. In: Cognitive Information Processing Workshop, pp. 114–119 (2008)
7. Philipose, M., Fishkin, K.P., Perkowitz, M., Patterson, D.J., Fox, D., Kautz, H., Hahnel, D.: Inferring activities from interactions with objects. IEEE Pervasive Computing 3(4), 50–57 (2004)
8. Rabiner, L.R.: A tutorial on hidden Markov models and selected applications in speech recognition. Proceedings of the IEEE, 257–286 (1989)
9. Tapia, E.M., Intille, S.S., Larson, K.: Activity Recognition in the Home Using Simple and Ubiquitous Sensors. In: Ferscha, A., Mattern, F. (eds.) PERVASIVE 2004. LNCS, vol. 3001, pp. 158–175. Springer, Heidelberg (2004)
10. van Kasteren, T.L.M., Englebienne, G., Kröse, B.J.A.: Transferring Knowledge of Activity Recognition across Sensor Networks. In: Floréen, P., Krüger, A., Spasojevic, M. (eds.) Pervasive Computing. LNCS, vol. 6030, pp. 283–300. Springer, Heidelberg (2010)
11. van Kasteren, T., Noulas, A., Englebienne, G., Kröse, B.: Accurate activity recognition in a home setting. In: 10th International Conference on Ubiquitous Computing, pp. 1–9. ACM (2008)
12. Yin, J., Yang, Q., Pan, J.J.: Sensor-based abnormal human-activity detection. IEEE Transactions on Knowledge and Data Engineering 20(8), 1082–1090 (2008)

Dynamic Workflow Management for Context-Aware Systems

José María Fernández-de-Alba, Rubén Fuentes-Fernández, and Juan Pavón

Abstract. Context-aware systems acquire and use information from their environment to provide services to their clients. A meaningful use of this context requires understanding the activities taking place, which are frequently organized as workflows. Managing a workflow implies, having its activities and their connections defined: storing the state, recognizing the activities execution, evaluate the decision conditions, and making the necessary actions. In order to make workflows adaptable to an evolving environment, they have an abstract definition that is grounded to different runtime instances depending on the actual conditions. Thus, there is a mutual influence between context and workflows: the execution of workflows depends on the context and changes it. Our work addresses these needs with a context-aware architecture that supports the abstract definition and management of workflows. The architecture provides access and subscription mechanisms to the context information, and coordinates the information sharing among context-aware application components. The architecture is illustrated with a use case of path guidance.

1 Introduction

Context-Awareness is defined as the ability of a system to provide its users with services adapted to the *context*, which includes preferences, current tasks and other environmental circumstances as location, time, available resources or certain conditions. In general, *context* refers to any information relative to people, places or objects that is found relevant for certain application's operation [1]. This property is cardinal for Ambient Intelligence (AmI) applications [7], which need to quickly adapt themselves to these kind of dynamic changes.

The previous definition raises two questions: how the system acquires an updated representation of the context, and how it actually adapts services to that representa-

José María Fernández-de-Alba · Rubén Fuentes-Fernández · Juan Pavón
Facultad de Informática, Universidad Complutense de Madrid,
Avenida Complutense, s/n. 28040 Madrid, Spain
e-mail: {jmfernandezdealba, ruben, jpavon}@fdi.ucm.es

P. Novais et al. (Eds.): Ambient Intelligence - Software and Applications, AISC 153, pp. 181–188.
springerlink.com

tion. In order to address the first one, the runtime system has to manage the process of building and updating abstract context elements. This requires gathering information from concrete sources and processing it to provide more abstract and refined information. This process will depend on the available resources and the required treatment of the information. A way of specifying this treatment is thus necessary in context-aware systems. The second question relates to the requirements of the system and it depends on the concrete application being developed.

Existing solutions for this functionality present some limitations. On one hand, comprehensive solutions are usually organized around quite complex and rigid architectures, like uFlow [5] or Ranganathan's [6]. They pose strong requirements on the supported component models and their dynamic organization. On the other hand, more flexible solutions usually dismiss some functionality, typically the context-awareness or the management of the information flows. This is the case of CAWE [2]. More examples are discussed and compared in the related work section.

Our work addresses this problem with a component-based architecture. It proposes splitting the context model in abstraction layers. The architecture establishes how to coordinate the components in order to process that information, making use of automatic discovery and binding mechanisms. A component that needs some information from the context only needs to declare what it wants to observe, and the information flow is automatically established to fulfil the request. This information can also flow in the other direction: a component may declare that it wants to communicate information to change the context. High-level expressions can be evaluated from less abstract context values, facilitating context reasoning. Workflow management, i.e. detection, tracking and execution of workflows, is built upon the previous functionality. Workflows are represented as activity diagrams which work using the high-level expressions as data.

The description of the architecture presents several patterns, which are explained using a concrete application as basis. The objective of the application is to give vocal and written indications in a smart phone to guide a student through a faculty to his teacher, and after the meeting, guide the student back to exit. The teacher's location is inferred through the information of his timetable, assuming that his location corresponds to the classroom he is giving the lesson, or his office if he does not have any class at that time. The user's location is inferred using certain sensors in the smart phone.

The rest of the paper is organized as follows. Sections 2, 3 and 4 present the context-aware architecture. Section 5 discusses other approaches that deal with context and workflow management in systems. Finally, section 6 presents some conclusions and future work on the approach.

2 Architecture Overview

The proposed context-aware architecture is inspired on distributed blackboard models [3]. It conceives a system as a set of interconnected *environments*, each one with its own *context container* (i.e. the blackboard) and a set of components working on it (i.e. the experts). The overall system is the result of the federation of these environments, which support collaboration through shared information.

An *environment* is a space (physical, virtual or hybrid) characterized by its context, which is composed by pieces of information with a semantic relationship. A specific component (the *Context* component) stores this information, and distributed expert components use and update it. Although many of the transformations consist on abstractions, there are also others that imply reasoning processes to drive actions towards a concrete goal, e.g. deciding the action for an actuator or producing new goals. These processes are required for workflow management: the abstract definition of the workflow and the semantic of its activities according to the current state of the context determine goals that the system is pursuing and potential actions that could fulfil them.

The previous organization is implemented with functionality organized in two layers (figure 1): *Context Management* and *Workflow Management*. The second layer is supported by the first one, and both of them are built on top of existing state-of-the-art component-oriented middleware. The main requirements for this middleware is that it has to provide support for the dynamic boot of components, the management of their dependencies and bindings, and service discovering. Web Services (possibly stateful) and OSGi are examples of suitable platforms. The following sections describe the basics of first layer necessary to understand the second layer.

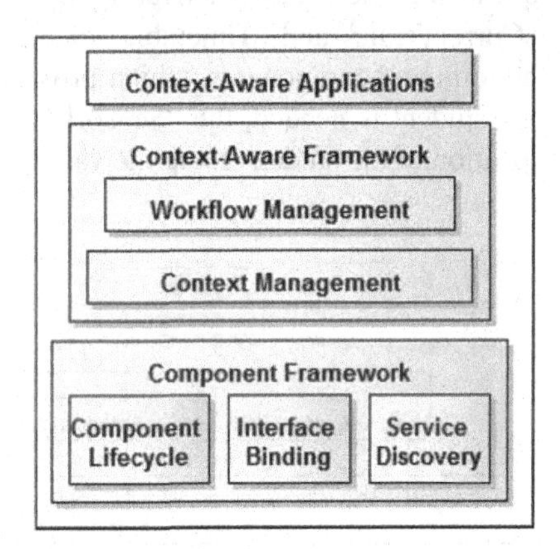

Fig. 1 Architecture organization in layers.

3 Context Management

This layer allows client components to request context elements and to evaluate expressions on their values. This layer consists of elements of the following types:

- **Context.** It represents an abstract container that acts as a *blackboard* [3]. It stores and makes visible every *ContextElement* to the components in the system.
- **Entity.** It represents an element that has a set of *Attributes* of different types. An example of entity is "Robert", a user of the system.

- **Attribute.** It represents an occurrence of an attribute on certain *Entity*. For example "location of Robert".
- **ContextElement.** It is the abstraction of a piece of information that belong to the context (*Entity* or *Attribute*). It maintains a set of *ContextObserver*s, that are and notified when it undergoes any change.
- **ContextObserver.** It represents a component that changes or receives notifications about changes in some *ContextElement*. It is also informed about newly created or destroyed *ContextElement*s.

This context management provides mechanisms for the evaluation of expressions on *ContextElement*s. It allows a *ContextObserver* to request the value of an *Attribute* to the *Context*. This value is obtained in a context-dependent way. If the element was previously requested, then it is directly provided; if not, the *Context* creates and publishes the instance, so that components suitable to handle that *Attribute* are made aware of the creation, triggering the calculation, which may in turn need successive requests to the context, until last components can provide the values directly (the sensors). When designing the components, the designer have to decide the granularity of the abstraction steps. This process is called *bottom-up* evaluation, because the value of the *ContextElement* come from the less abstract components.

Figure 2 (left) shows how the attribute "Location" for the mentioned application is calculated. When the application needs the concrete location of the teacher, it requests it to the context, activating the "TimetableLocationUpdater", which requests "CurrentTime" and "Timetable" of the teacher. They are in turn calculated by the bottommost components, which provide the values directly. When the location of the student is needed, the "SensorLocationUpdater" is activated, which infers the location of the student using the value returned by certain sensor.

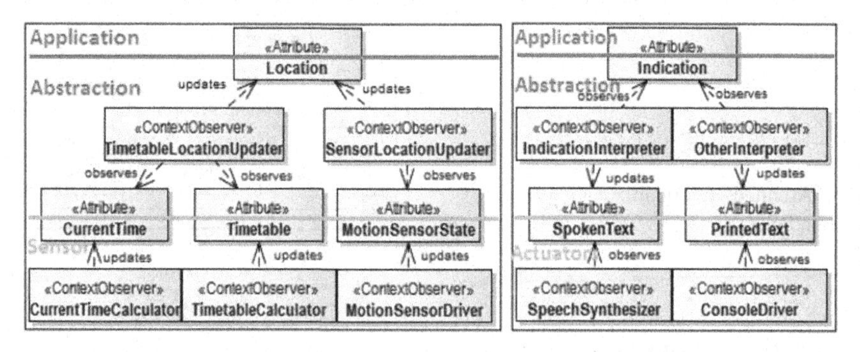

Fig. 2 Bottom-up and top-down evaluations.

The *top-down* evaluation process works in the opposite direction. A component can request a certain *Attribute* for change. If the instance did not exist, the *Context* creates and publishes it, and the components that can use its value are made aware, subscribing themselves to its changes. That causes the successive request for change of other *Attribute*s in cascade, until last components can use the value directly (the actuators). Then, each time the value of original *Attribute* is changed, the changes are notified downwards, ultimately activating the actuators.

Figure 2 (right) shows how changing “Indication” causes the triggering of two different interpreters, which in turn update the attributes that are directly used by the actuators: those that contain the text that should be spoken and the text that should be printed.

For workflow management, is more convenient to resolve the evaluation of some *Attributes* not only matching its type with the available components, but also considering other conditions. For this reason, the architecture considers *parametrized evaluators*, which act like *ContextObserver* factories. These *ContextObservers* are able to select an implementation for the evaluation depending on the context runtime conditions, and to create a new instance of it to calculate the value. This pattern is used as the basis for component behaviour selection, which permits the implementation of the *Workflow Management* pattern, as explained below.

4 Workflow Management

This layer deals with the context-aware support for workflow execution. Workflows are described as networks of activities in activity diagrams. The management of a workflow implies maintaining its state, and read or manipulate the context when necessary while the activities are executed in the defined order. This is done performing bottom-up or top-down evaluations. This way, if completion of certain activity is determined by certain context condition, e.g., an activity involving a user that ends when his *Location* attribute takes certain value, a bottom-up evaluation is done in order to examine the *Location* value, which is obtained from sensors. Similarly, if another activity implies certain abstract action, e.g., transmit an indication, a top-down evaluation is done, since the workflow modifies the *Indication* attribute of the *environment*, and the modification goes down to the actual components that handle the change, like a loudspeaker driver.

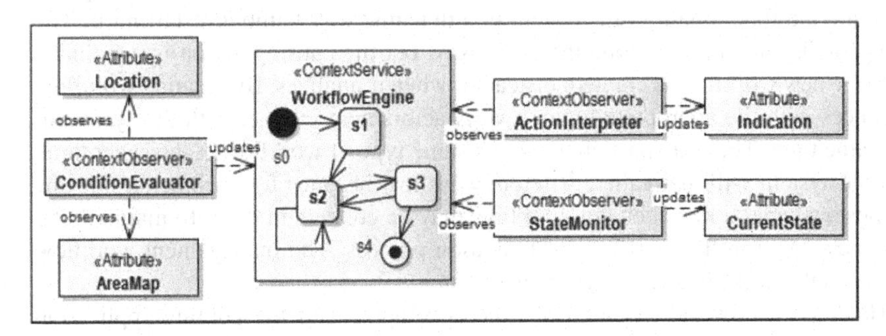

Fig. 3 Workflow pattern structure.

The architecture supports the previous functionality with the pattern in figure 3, which consists of the following elements:

- **Condition Evaluator.** The pattern defines a workflow as an activity diagram in which activities may read the context. This component asks the *Context*, generating bottom-up evaluations. Then, the results of this evaluation are passed the workflow engine for reading.
- **Workflow Engine.** This component stores the abstract definition of the workflow and manages its current state. It receives information from the *Condition Evaluator* to determine if certain activity has been completed, or to choose the correct transitions in the decision nodes. Also, according to the activity to be executed, it determines the output to be sent to the *Action Interpreter*.
- **Action Interpreter.** It receives abstract actions from the *Workflow Engine*. These actions indicate context changes to be forwarded. These changes are performed as top-down evaluations, so the rest of interested components in the system can be notified.
- **State Monitor.** The workflow itself can be regarded as an *Entity* where the "currentState" attribute is interesting for certain components. This component monitors the state of the workflow asking the *Workflow Engine*, and is able to provide the mentioned *Attribute*, if asked. This allows any potential client component to know the current state of the workflow.

This structure promotes isolating in different components the domain vocabulary of the context and the workflow logic, facilitating reuse. It also provides means to implement the workflow actions and condition evaluation as context-aware elements. Finally, this pattern is the basis to support nested workflows (an activity that is implemented as another workflow) and workflow instantiation (an activity that creates an instance of another workflow which becomes independent).

Nested activities offer the possibility of further decomposing an activity in others. This occurs when the component that determine the conclusion of certain activity implements in turn another workflow pattern, so that the conclusion depends on the finalization of the last workflow. These nested activities allow extending and adapting the functionality of a system in a dynamic way without restarting it. The workflow instantiation is done the same way, but the causing action is concluded after the new workflow is created, instead of when it finalizes. By creating workflow instances with the appropiate logic, several actors may interact with the system at the same time. They can participate in the same type of workflow at the same time, and the system will maintain a different workflow instance for each actor. For this purpose, a general management workflow may be created, in order to maintain the bussiness logic in the workflow instantiation process. The management workflow that does this task is called a *workflow initiator*.

Having the previous mechanisms, the development of the guiding application mentioned during the description consists in two basic tasks: specifying the activity diagram with its conditions and actions, and the necessary *ContextObservers* to handle that conditions and actions.

5 Related Work

The problem of the support to manage context-aware workflows has been already considered in some works. Most of them may provide an infrastructure or middleware to support the storage, distribution and management of information, but few describe how the components behave against system configuration changes, i.e., how the system actually adapts itself to changes in its parts. This constitutes the main advantage of this work compared with them.

Some of these solutions make use of external context management services and encapsulate them for their use. For instance, in uFlow [5] and Ranganathan's [6] are proposed solutions in this line. This approach limits the level of control a workflow can make over the different devices and components on the system, since it need to rely on the same predefined interfaces. Although this is not usually a problem, it provides a lower level of control for the specific details in workflow definition. Similarly, the Context4BPEL [8] defines a static layered model, that limits the granularity of the abstraction levels that can be designed. In our architecture, the information flows may be implemented dividing them at any desired levels of abstraction.

Talking about the workflow definition, there are multiple alternatives. In CAWE [2] and uFlow [4] are discussed some of their own, while Ranganathan [6] proposes using BPEL, which is a standard language for the definition of business processes. The advantage on the first two works is that they include context-dependent primitives, while the third does not. However, the third alternative uses a well-known and defined language. Our work uses a workflow definition structure that allows using any workflow engine and language, as the engine is wrapped with modules that adapt its inputs and outputs.

6 Conclusions

This paper has introduced a generic architecture for context-aware workflow management. Its definition includes patterns to develop subsystems that are able to track context conditions and to perform different actions on it according to activity diagrams.

The architecture has the following advantages: first, the generation of information elements is made when the situation requires it (also called opportunistic reasoning), which saves resources if there is no application component interested in a certain context element. Second, it supports the dynamic reconfiguration of the system when available components change by taking care of new bindings between components and context elements. And third, the resulting system is dynamically extensible, i.e., its functionality can be extended or adapted at runtime without restarting or recompiling any component, and permits deciding the level of granularity in the abstraction levels. That is because adding functionality consist only in including new components observing the context. Adaptation may be done substituting one of these components. These features, especially the last two, leverage the possibilities when defining abstract context-dependent workflows.

The presented work is part of a wider effort to provide a general architecture and infrastructure for AmI applications. As future work, this architecture will include a third layer, the *Mobility Management*. Using these patterns it would be possible to create distributed environments, in which an activity can be defined to involve many spaces. For this, it would be necessary to coordinate them to exchange the adequate information, and to maintain the consistency. Also, it is planned to introduce management for real time constraints in the environments.

Acknowledgements. The authors acknowledge support from the project *Agent-based Modelling and Simulation of Complex Social Systems (SiCoSSys)*, supported by Spanish Council for Science and Innovation, with grant TIN2008-06464-C03-01. Also, we acknowledge support from the Programa de Creación y Consolidación de Grupos de Investigación UCM-BSCH GR35/10-A.

References

1. Abowd, G., Dey, A., Brown, P., Davies, N., Smith, M., Steggles, P.: Towards a Better Understanding of Context and Context-Awareness. In: Gellersen, H.-W. (ed.) HUC 1999. LNCS, vol. 1707, pp. 304–307. Springer, Heidelberg (1999)
2. Ardissono, L., Furnari, R., Goy, A., Petrone, G., Segnan, M.: Context-Aware Workflow Management. In: Baresi, L., Fraternali, P., Houben, G.-J. (eds.) ICWE 2007. LNCS, vol. 4607, pp. 47–52. Springer, Heidelberg (2007)
3. Corkill, D.: Blackboard Systems. AI Expert 6(9) (1991)
4. Han, J., Cho, Y., Choi, J.-Y.: Context-Aware Workflow Language Based on Web Services for Ubiquitous Computing. In: Gervasi, O., Gavrilova, M.L., Kumar, V., Laganá, A., Lee, H.P., Mun, Y., Taniar, D., Tan, C.J.K. (eds.) ICCSA 2005, Part II. LNCS, vol. 3481, pp. 1008–1017. Springer, Heidelberg (2005)
5. Han, J., Cho, Y., Kim, E., Choi, J.-Y.: A Ubiquitous Workflow Service Framework. In: Gavrilova, M.L., Gervasi, O., Kumar, V., Tan, C.J.K., Taniar, D., Laganá, A., Mun, Y., Choo, H. (eds.) ICCSA 2006, Part IV. LNCS, vol. 3983, pp. 30–39. Springer, Heidelberg (2006)
6. Ranganathan, A., Campbell, R.H.: A Middleware for Context-Aware Agents in Ubiquitous Computing Environments. In: Endler, M., Schmidt, D.C. (eds.) Middleware 2003. LNCS, vol. 2672, pp. 143–161. Springer, Heidelberg (2003)
7. Remagnino, P., Hagras, H., Monekosso, N., Velastin, S.: Ambient Intelligence – a gentle introduction. In: Remagnino, P., Foresti, G., Ellis, T. (eds.) Ambient Intelligence, pp. 1–14. Springer, New York (2005)
8. Wieland, M., Kaczmarczyk, P., Nicklas, D.: Context integration for smart workflows. In: Proceedings of the 2008 Sixth Annual IEEE International Conference on Pervasive Computing and Communications, pp. 239–242. IEEE Computer Society, Washington, DC (2008)

Part II
Short Papers

Robotic UBIquitous COgnitive Network

Giuseppe Amato, Mathias Broxvall, Stefano Chessa, Mauro Dragone, Claudio Gennaro, Rafa López, Liam Maguire, T. Martin Mcginnity, Alessio Micheli, Arantxa Renteria, Gregory M.P. O'Hare, and Federico Pecora

Abstract. Robotic ecologies are networks of heterogeneous robotic devices pervasively embedded in everyday environments, where they cooperate to perform complex tasks. While their potential makes them increasingly popular, one fundamental problem is how to make them self-adaptive, so as to reduce the amount of preparation, pre-programming and human supervision that they require in real world applications. The EU FP7 project RUBICON develops self-sustaining learning solutions yielding cheaper, adaptive and efficient coordination of robotic ecologies. The approach we pursue builds upon a unique combination of methods from cognitive robotics, agent control systems, wireless sensor networks and machine learning. This paper briefly illustrates how these techniques are being extended, integrated, and applied to AAL applications.

1 Introduction

Building smart environments out of multiple robotic devices extends the type of application that can be considered, reduces their complexity, and enhances the

Giuseppe Amato · Claudio Gennaro
ISTI-CNR

Mathias Broxvall · Federico Pecora
Örebro Universitet

Stefano Chessa · Alessio Micheli
Università di Pisa

Mauro Dragone · Gregory M.P. O'Hare
University College Dublin,

Rafa López
Robotnik Automation

Liam Maguire · T. Martin Mcginnity
University of Ulster

Arantxa Renteria
Tecnalia
e-mail: coordinator@fp7rubicon.eu

P. Novais et al. (Eds.): Ambient Intelligence - Software and Applications, AISC 153, pp. 191–195.
springerlink.com

individual values of the devices involved by enabling new services that cannot be performed by any device by itself. Consider for instance the case of an automatic vacuum cleaner that avoids cleaning when any of the inhabitants are home after receiving information from the home alarm system.

Current robotic ecologies [1] strictly rely on models of the environment and of its associated dynamics. For instance, in AAL settings, they require pre-defined models of both the activities of the user they try to assist and the services that should be carried out to assist them. Crucially, they lack the ability to proactively and smoothly adapt to evolving situations and to subtle changes in user's habits and preferences. All of these limitations make such systems difficult to deploy in real world applications, as they are tailored to the specific environments, hardware configurations, applications and users, and they can soon become unmanageablely complex.

Multi-agent and robotics applications have often relied on machine learning solutions to free the developer from having to specify the details and consequences of the interaction between each agent and its environment, and to deal with noisy and uncertain sensor data. However, until now, strict computational constraints have posed a major obstacle to translating the full potential benefits of these results in robotic ecologies. Furthermore, even when they are successfully applied, they usually require expensive training sessions and costly human supervision to drive each adaptation step.

The EU FP7 project RUBICON (Robotic UBIquitous COgnitive Network) builds on existing solutions to develop the concept of self-sustaining learning for robotic ecologies. Specifically, RUBICON investigates how all the participants in a robotic ecology can cooperate in using their past experience to improve their performance by autonomously and proactively adjusting their behaviour and perception capabilities in response to a changing environment and user's needs.

2 The RUBICON Approach

In order to test and develop RUBICON, we use a smart home test-bed laboratory - a fully functional apartment of over $40m^2$ equipped with automated doors and blinds, and sensors such as gas/water/smoke/movement detectors and microphones. In addition, the test-bed is being extended with mobile robots and with wireless sensor networks (WSN) nodes, each comprising a computational unit with radio and sensors.

The resulting RUBICON ecology is required to:

- Assist the users in their daily living and also alert relevant stakeholders of potentially dangerous or anomalous behaviour and/or situations.
- Learn to leverage and enhance its context awareness and reasoning abilities (e.g. including the ability to recognize user's activities, to locate humans and/or robots as well as dangerous situations);
- Adapt and tune its abilities to the characteristics of the environment where it is deployed and to the behaviour of its inhabitants, for instance, adapting to the

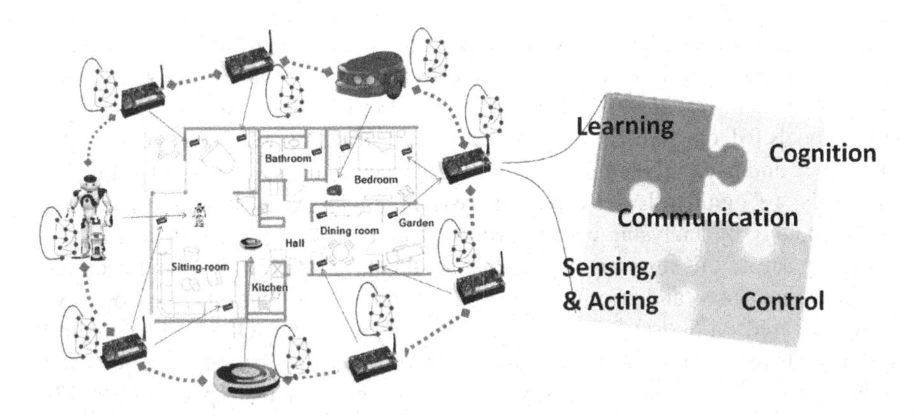

Fig. 1 Left: System Diagram, Right: RUBICON Layered Architecture.

fitting of new furniture and new carpets, and to changes in the user's preferences and needs.

- Exhibit robustness, reliability and graceful degradation of performance when some of its devices are removed or replaced, as well as the ability to seamlessly incorporate additional ones.

The RUBICON layered architecture, illustrated in Fig. 1, builds upon the PEIS middleware [1] to provide a de-centralized Communication Layer for collaboration between functional processes (such as those used to classify household sounds, localize users and/or robots, as well as navigation and other robotic skills) running on separate and heterogeneous devices. In order to enable simple and effective access to the transducer and actuator hardware on wireless nodes, all of these resources are abstracted and accessed as communication channels built over the MadWise Stream System framework [2].

In order to achieve necessary and meaningful tasks in a goal-oriented, coordinated fashion, the nodes participating in a RUBICON system are controlled by a Control Layer built on agent [3] and timeline-based planning [4] technologies. In this manner, a robotic ecology is capable of finding and using alternative means to accomplish its goals when multiple courses of action/configuration are available. For instance, a robot may decide to localize itself with its on-board sensors, or to avail itself of the more accurate location information from an environmental camera.

The key to enabling adaptive behaviour is to learn to extract meaning from noisy and imprecise sensed data, and also to learn what goals to pursue, and how to pursue them, from experience, rather than by relying on pre-defined strategies.

These issues are tackled by exploiting a distributed learning infrastructure and *flexible environmental memory* - the RUBICON Learning Layer. In particular, the Learning Layer is used to: (i) provide predictions which depend on the temporal history of the input signals (e.g. the users future location [5], or the probability of success of performing an action or using a device in a given situation), and (ii) analyze, process and fuse sensed information to extract refined goal-significant information (e.g. recognize that the user is cooking by monitoring its location as well as the sensors signalling when kitchen appliances are in use). To these ends, we make use of *Recurrent Neural Networks* (RNN), and specifically of *Reservoir Computing* (RC) models [6], due to their modular, networked structure, which can be naturally distributed and overlaid on top of the RUBICON ecology, as represented in Fig. 1.

Finally, a Cognitive Layer drives the reasoning and self-sustaining capabilities of the ecology by analysing current events, as determined by the Learning Layer, reasoning across this information and historical data and behaviours, and deciding on appropriate goals and priorities for the Control Layer. To this end, the Cognitive Layer is based on *Self-Organizing Fuzzy Neural Network* (SOFNN) [7] - hybrid systems where neural networks are used to learn fuzzy membership functions and create fuzzy rules that may be easily interpreted. The particular appeal of SOFFN is their capacity for self-organizing structural growth through the addition and the pruning of neurons driven by novelty detection and habituation mechanisms [8].

3 Conclusion and Future Work

We believe that the extension and integration of the techniques we discussed along the lines illustrated in this paper promises to solve many of the problems that still obstruct the implementation and diffusion of smart robotic environments outside research laboratories. Future work within the project RUBICON will refine and implement RUBICONs high-level architecture and validate it in realistic settings.

Acknowledgements. This work is partially supported by the EU FP7 RUBICON project (contract n. 269914) - www.fp7rubicon.eu.

References

1. Broxvall, M., Seo, B.S., Kwon, W.Y.: The PEIS Kernel: A Middleware for Ubiquitous Robotics. In: Proc. of the IROS 2007 Workshop on Ubiquitous Robotic Space Design and Applications, San Diego, California (October 2007)
2. Amato, G., Chessa, S., Vairo, C.: MaDWiSe: A Distributed Stream Management System for Wireless Sensor Networks. Software Practice & Experience 40(5) (2010)
3. Muldoon, C., O Hare, G.M.P., O' Grady, M.J.: AFME: An Agent Platform for Resource Constrained Devices. In: Proceedings of the ESAW 2006 (2006)
4. Pecora, F., Cirillo, M.: A Constraint-Based Approach for Plan Management in Intelligent Environments. In: Proc. of the Scheduling and Planning Applications Workshop at ICAPS 2009 (2009)
5. Bacciu, D., Gallicchio, C., Micheli, A., Chessa, S., Barsocchi, P.: Predicting User Movements in Heterogeneous Indoor Environments by Reservoir Computing. In: Proc. of the IJCAI Workshop on Space, Time and Ambient Intelligence (STAMI), Barcellona, Spain, pp. 1–6 (2011)
6. Lukosevicius, M., Jaeger, H.: Reservoir computing approaches to recurrent neural network training. Computer Science Review 3(3), 127–149 (2009)

7. Prasad, G., Leng, G., McGinnity, T.M.: On-line identification of self-organising fuzzy neural networks for modelling time-varying complex systems. In: Plamen, et al. (eds.) Evolving Intelligent Systems: Methodology and Applications, pp. 302–324. Wiley-IEEE Press (2010)
8. Mannella, F., Mirolli, M., Baldassarre, G.: Brain Mechanisms underlying Learning of Habits and Goal-Driven Behaviour: A Computational Model of Devaluation Experiments Tested with a Simulated Rat. In: Tosh, C. (ed.) Neural Network Models. Cambridge University Press

A U-HealthCare System for Home Monitoring

Giovanna Sannino and Giuseppe De Pietro

Abstract. This paper presents an advanced ubiquitous system for home health monitoring. The main goal of the research presented in this paper is to develop a user-friendly and context-aware system that uses a rule-based Decision Support System to elaborate the data captured by the sensors. The paper also describes a case study where important benefits for patients have been revealed thanks to the use of the proposed home health monitoring system.

1 Introduction and Related Works

Recently wireless sensor technology has been riveting a lot of attention as a possible means to realize the ubiquitous computing environment in everyday life. There has been much research on the area, such as ParcTab [1] and ActiveBadge [2] as two pioneers, and many relevant areas have got rapid development, including context-aware computing [3-5]. In the context-aware computing paradigm the context is any information that can be used to characterize the situation of an entity which may be a person, a place, or an object that is considered relevant, including the user and the application themselves.

The application of context-aware computing in the field of ubiquitous computing continues to be of interest to many researchers. It creates the required smart features that allow flexible interaction between the user and the environment. Tele-Monitoring at home is one of major trends that impacts worldwide especially in the developed countries, contrasting increase in health care costs. The availability of a smart living environment, in the form of a smart home, will help alleviate some aspects of this problem and hence release for other patients resources that would otherwise go to long-term health support of an individual patient, will enable early diagnosis of chronic health conditions, and will make clinical visits more efficient due to the availability of objective information prior to such visits [6,7].

Giovanna Sannino · Giuseppe De Pietro
Institute of High Performance Computing and Networking - National Research Council of Italy

P. Novais et al. (Eds.): Ambient Intelligence - Software and Applications, AISC 153, pp. 197–200.
springerlink.com

In this paper we present an advanced ubiquitous system for home health monitoring with a rule-based DSS able to enhance the accuracy of potentially dangerous Heart Rate variability by taking into account patient context information, like postures, movements or falls, and other information like room temperature.

2 Software Architecture

The proposed advanced mobile system is based on the architecture presented in [8]. The modular software architecture is devised to simply accommodate technological or functional changes, like the introduction of new hardware or new system software, and is divided into three different layers:

1. The Data Layer provides user interfaces and mechanisms to manage sensors data and patient information. It collects information about the patient monitored and data from monitoring devices and calculates more complex parameters such as the peak of QRS or the peak of plethysmography.
2. The Decisional Layer elaborates data coming from the Data Layer, so recognizing critical situations and determining the most appropriate actions to be performed. To this aim, a rule-based decision support system has been designed and implemented. The used rules represent the expert's knowledge formalization of anomalies that should be detected, and the consequent actions to take.
3. Finally, the Action Layer executes the actions inferred by the Decisional Layer by implementing mechanisms to produce reactions like the generation of alarms and messages or starts the video streaming by IP camera.

The improvement here introduced, compared with [8], is the addition of three new modules in the Data Layer. Now, the architecture provides a software module dedicated to the fall detection, a module for IP Camera communication, and a module dedicated to communication with temperature sensor.

3 Software and Hardware Components

All software layers are implemented for resource-limited mobile devices, such as PDA and smart phone, using the java programming language. According to the architectural model described in [8], the software components implemented are organized on three levels. For each module of each level at least one component was expected, and if necessary more than one have been provided.

To verify the system performance we conducted a number of tests using the following set of hardware devices. ECG Monitor and Accelerometer are part of the Alive Heart Monitor™ sensor, which is a wireless health monitoring system. The monitor uses wireless Bluetooth networks to transmit ECG and accelerometer data in real time. The Alive Pulse Oximeter Monitor™ is a wearable medical device that reads oxygen saturation on the finger or earlobe, and transmits the data in real time via Bluetooth. The IP camera used is an AXIS 210A IP Network Camera™, a professional network camera for indoor monitoring over IP networks.

The ZED-THL-M is a ZigBee device that records temperature, humidity and light measurements and transmits the data at regular intervals to a ZB-Connection Gateway. Finally, we used the Datrend AMPS-1 to test the overall system. It is a hardware compact patient simulator that can reproduce 52 different kinds of arrhythmia, so simulating many potentially dangerous situations for testing purposes. For tests we also used a smart phone, Nokia 5800, equipped with Symbian Operating System.

4 A Case Study

With heart failure, simultaneous monitoring of cardiac and other physiologic parameters while patients perform normal daily activities can provide a safety net for documenting cardiac events and may be useful for health care providers to improve management of the disease. Furthermore, home-based telemonitoring decreases emergency room visits and hospitalizations when compared to standard care [9]. The proposed case study aims to monitor physiological parameters of patients who suffer from health failures – such as cardiomyopathies, blockages or congestive heart failure – and to detect patient falls. Namely, such patients wear sensors and are monitored within their homes. Their ECG signal is caught by sensors and preprocessed by a monitoring station on mobile devices, which also alerts medical staff in case of emergency or notifies patient in case of an anomalous, but not serious, situation. Moreover, a fall detection system has been realized to detect faints through 3-axes accelerometer signal, and some video surveillance cameras are activated whenever the system detects a fall. Finally, the system uses some temperature sensors to monitor room state. Figure 1 shows a possible scenario.

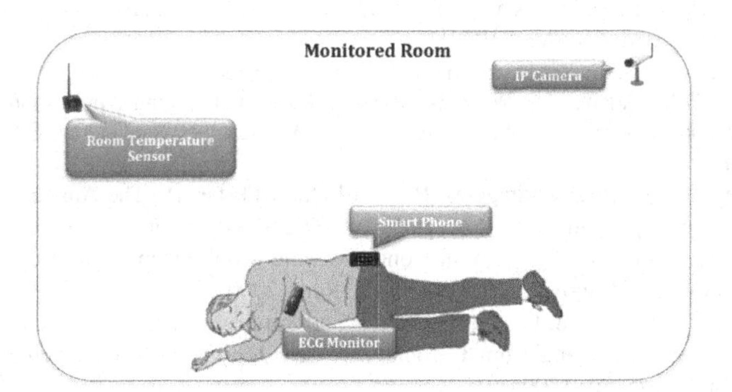

Fig. 1. U-HealthCare System - Home Health Monitoring – Scenario

In the scenario, while a patient performs normal daily activities, she/he falls. The system, through the ECG monitor checking her/his heart rate, detects in real time the value of heart rate and by the 3-axes accelerometer data calculates her/his posture, her/his movement, and if the patient has fallen down. Contemporaneously the system captures the data from the temperature sensor of the room the patient stays in. The system detects patient's fall, so it immediately sends an alarm message to the emergency station and/or to her/his doctor and starts the video streaming by IP camera to allow medical personnel to monitor the state of the patient.

The system also monitors the room temperature to decide to take some actions because for patients in our case study – elderly or people with chronic heart problems –room temperature should be in a given range. In fact a too high temperature can cause myocardial infarction and/or stroke as well as a heart rate increase followed by a blood-pressure relieving and fainting. Conversely, a too low temperature can cause high blood pressure, a risk factor for chronic diseases and stroke; when the temperature is below 4 °C, heart patients' risk is estimated as doubled.

5 Conclusions

This paper has described an advanced system for home health monitoring. The system uses a rule-based Decision Support System to elaborate the captured sensors data and to achieve more reliable alarm and warning generation. The paper has also presented a case study with a scenario to show the utility and the benefits of the system. The scenario shows the feasibility of a context-aware approach for health monitoring, able to diminish the generation of unnecessary alarms. Future work will aim at both implementing new services and enhancing the existing ones.

References

[1] The ParcTab Ubiquitous Computing Experiment (1998), `http://www.ubiq.com/parctab/csl9501/paper.html`

[2] The Active Badge System (1992), `http://www.uk.research.att.com/ab.html`

[3] Schilit, B.N., Adams, N., Want, R.: Context-Aware Computing Applications. In: Proc. Workshop on Mobile Computing Systems and Applications, Santa Cruz, CA, pp. 85–89 (1994)

[4] Harter, A., Hopper, A., Steggles, P., Ward, A., Webster, P.: The Anatomy of a Context-Aware Application. In: Proc. of the ACM/IEEE MobiCom, pp. 59–68 (1999)

[5] Chen, G., Kotz, D.: A Survey of Context-Aware Mobile Computing Research. Dartmouth College, Hanover, NH, USA. Tech. Rep. (2000)

[6] Botsis, T., Hartvigsen, G.: Current Status and Future Perspectives in Telecare for Elderly People Suffering from Chronic Diseases. Journal of Telemedicine and Telecare 14(4), 195–203 (2008)

[7] Friedewald, M., Da Costa, O., Punie, Y., Alahuhta, P., Heinonen, S.: Perspectives of ambient intelligence in the home environment. Telematics and Informatics 22, 221–238 (2005)

[8] Sannino, G., De Pietro, G.: An Intelligent Mobile System For Cardiac Monitoring. In: Proc. of IEEE Healthcom 2010, Lyon, France, pp. 52–57 (2010)

[9] Cordisco, M.E., Benjaminovitz, A., Hammond, K., Mancini, D.: Use of telemonitoring to decrease the rate of hospitalization in patients with severe congestive heart failure. American Journal of Cardiology 84(7), 860–862 (1999)

Useful Research Tools for Human Behaviour Understanding in the Context of Ambient Assisted Living

Pau Climent-Pérez, Alexandros Andre Chaaraoui, and Francisco Flórez-Revuelta

Abstract. When novice researchers in the fields of Computer Vision and Human Behaviour Analysis/Understanding (HBA/HBU) initiate new projects applied to Ambient-Assisted Living (AAL) scenarios, a lack of specific, publicly available frameworks, tools and datasets is perceived. This work is an attempt to fill that particular gap, by presenting different field-related datasets—or benchmarks—, according to a taxonomy (which is also presented), and taking into account their availability as well as their relevance. Furthermore, it reviews and puts together a series of tools—either frameworks or pieces of software—that are at hand (although dispersed), which can ease the task. To end with the work, some conclusions are drawn about the reviewed tools, putting special emphasis in their generality and reliability.

1 Reviewed Datasets and Tools

In this paper, a taxonomy is followed, which is based on others seen in the literature such as [7, 9]. Under this taxonomy (Table 1), 'actions' are classified into increasing degrees of semantic richness (DoS) and the time involved in the analysis.

According to the presented degrees of semantics, and having ADL recognition and AAL as targets, the following datasets stand out:

KTH human motion dataset [10]: This action database contains six types of human actions performed by 25 subjects in four different scenarios. These are performed in over 2 000 sequences. Backgrounds are homogeneous and free of clutter. Video files are classified by actions, so that unwanted actions can be excluded easily.

Weizmann human action dataset [4]: Gorelick et al. use static front-side cameras to record single human motion from 9 subjects in different environments. About 340 MB of video sequences are available. The corresponding background sequences, with no subjects, and the subtraction masks—either with post-aligning or without it—are available too.

Pau Climent-Pérez · Alexandros Andre Chaaraoui · Francisco Flórez-Revuelta
Department of Computing Technology, University of Alicante, Ctra. San Vicente del Raspeig, s/n, 03690 San Vicente del Raspeig, Alicante, Spain
e-mail: {pcliment,alexandros,florez}@dtic.ua.es

P. Novais et al. (Eds.): Ambient Intelligence - Software and Applications, AISC 153, pp. 201–205.
springerlink.com

Table 1 Classification of tasks according to the degree of semantics (DoS) involved

DoS	Time lapse	Description
Motion	frames, seconds	Movement detection, Background subtraction and Segmentation; Gaze and Head-pose estimation.
Action	seconds, minutes	Person–object interaction. Recognise simple human primitives (sitting, standing, walking, etc.)
Activity	minutes, hours	Sequences of ordered actions. ADLs[a] are recognised (e.g. cooking, taking a shower or making the bed).
Behaviour	hours, days, ...	Highly-semantic comprehension (ways of living, personal habits, routines of ADLs)

[a] ADLs stands for 'Activities of Daily Living'.

INRIA Xmas motion acquisition sequences [13]: In this dataset, 11 actors perform 13 actions. These actions are performed three times each, in an arbitrary chosen angle in relation to the view-point. Backgrounds and illumination settings are static and free of clutter.

TUM kitchen dataset [12]: This dataset targets ADLs in a kitchen scenario at the action level. Video; marker-less full-body tracking; RFID tag readings from fixed readers at the placemat, the napkin, the plate and the cup; and sensor data from magnetic sensors at doors and drawers are available.

MuHAVI dataset [11]: This dataset includes video data obtained from multiple cameras. Images are taken with night street light illumination at a constant but uneven background. At each corner and each side of a rectangular platform a camera is installed. These capture 16 different composite actions and one highly complex activity, performed by 7 actors, three times each.

CAVIAR test scenarios [3]: Here, images are taken at two different scenarios: an entrance lobby and a shopping centre. Ground-truth data is provided in XML format at frame level. Video sequences, taken from wide angle cameras installed as surveillance cameras at the ceiling corners, include several persons, as well as crowd movements.

CMU-MMAC database [2]: This dataset targets cooking and food preparation activities. Videos come along with audio and motion data. Five subjects are shown in a kitchen while preparing five different recipes.

PlaceLab datasets [6]: The PlaceLab[1] live-in laboratory provides a full home-like environment for all sorts of data gathering, useful for AAL scenarios among others. A new version (PLIA2) improves data sharing and visualization by employing new formats. PLIA2 includes 4 hours of video data (infrared and RGB). Accelerometer data is recorded by *MITes*, which are attached to objects such as remote controls, chairs, etc.

Table 2 evaluates the datasets with respect to the most relevant properties, which have been chosen having in mind possible constraints in HBA methods.

In what is related to tools, only recently, generalistic, interoperation-enabling approaches have been published. Some of these first steps in multipurpose design of tools; such as languages, meta-models and frameworks are presented:

[1] `http://architecture.mit.edu/house_n/data/PlaceLab/PlaceLab.htm`

Table 2 Comparison of dataset features

Dataset	DoS	'Actions'	Multi-view	Maximum resolution	Background type	Silhouettes	Out-/Indoor
KTH	actions	6	No	160×120	simple	No	both
Weizmann	actions	10	No	180×144	simple	Yes	outdoor
INRIA-XMAS	actions	13	Yes	390×291	simple	Yes	indoor
TUM Kitchen	actions	10[a]	Yes	780×582	simple	No	indoor
MuHAVI	*both*	17	Yes	720×576	complex	Yes[b]	indoor
CAVIAR	activities	6	Yes	384×288	complex	No	indoor
CMU-MMAC	activities	5	Yes	1024×768	simple	No	indoor
PlaceLab (PLIA2)	activities	6	Yes	320×240	simple	No	indoor

[a] Approximately 10 annotated sub-actions of 1 activity: setting the table.
[b] They are provided in the Manually-Annotated Subset (MAS).

Home markup language [8]: HomeML is an XML based schema for representation of information within smart homes. As data taken at a smart home scenario belongs to heterogeneous nature, and is captured by different type of sensors; this language offers an open standard for the exchange of data in a system-, application- and format-independent way. HomeML supports a data structure which is designed upon the most used standards in integration of home services and devices: OSGi and KNX.

ViPER – The Video Performance Evaluation Resource[2]: ViPER is a framework which targets semantic video analysis and includes several tools which make system evaluation easier. As such, the framework includes a Ground Truth Authoring Tool which includes a GUI to edit ground truth data and check generated metadata frame by frame.

Hong et al.'s activity recognition meta-model: In [5], Hong et al. present a new meta-model for activity recognition in smart homes. A diagram, which is similar to an Entity-Relation, is used to build evidential networks which express the interaction between recognised activities and objects. This way, relationships between activities, objects and associated sensors, as well as generalization at activity level and compulsory or optional interaction with objects can be captured.

BehaviourScope Framework [1]: It is a framework for detailed behaviour interpretation of the elderly. Its aim is to process, communicate and present heterogeneous sensor data in an automated form, in order to infer high-level semantic data, which can be further processed in applications and services (generation of alarms, reports, triggers and answers to queries is considered).

OpenAAL [14]: OpenAAL has been developed since 2007, as it started as the technical development of the SOPRANO Integrated Project (6th Framework Programme of IST). On top of the OSGi service-oriented framework, OpenAAL provides generic platform services based on three main components: 1) A Context Manager, where sensors data and user inputs are collected and stored supporting context reasoning at multiple levels of abstraction; 2) A Procedural Manager, which is in charge of handling installation-independent workflows which are able to react to situations of interest; and 3) The Composer, which selects the available services in the concrete installation to achieve the abstract service goals; these are described in the installation-independent workflows.

[2] `http://viper-toolkit.sourceforge.net/`

2 Conclusions

This paper has covered the most used datasets in the field of HBA and AAL scenarios, classifying them according to a previously defined taxonomy. It has been observed that dataset properties vary widely in respect to the quality of their images, additional data and characteristics of the environment. Advantages and particular difficulties of each dataset have been pointed out in order to ease an election.

The performed analysis on the available frameworks and tools shows that few of these have been released for public use and most research projects of this kind are unfortunately not available or, sometimes, discontinued.

Finally, the presented frameworks and tools are helpful for speeding up repetitive development stages and, more importantly, to reach a common approach among researchers in the field.

Acknowledgements. This work has been partially supported by the Spanish Ministry of Science and Innovation under project "Sistema de visión para la monitorización de la actividad de la vida diaria en el hogar" (TIN2010-20510-C04-02).

References

1. Bamis, A., Lymberopoulos, D., Teixeira, T., Savvides, A.: The BehaviorScope framework for enabling ambient assisted living. Personal and Ubiquitous Computing 14(6), 473–487 (2010)
2. De la Torre, F., Hodgins, J., Montano, J., Valcarcel, S., Macey, J.: Guide to the Carnegie Mellon University Multimodal Activity (CMU-MMAC) Database. Tech. rep. (2009)
3. Fisher, R.: CAVIAR Test Case Scenarios (2007), `http://groups.inf.ed.ac.uk/vision/CAVIAR/CAVIARDATA1/`
4. Gorelick, L., Blank, M., Shechtman, E., Irani, M., Basri, R.: Actions as space-time shapes. IEEE Transactions on Pattern Analysis and Machine Intelligence 29(12), 2247–2253 (2007)
5. Hong, X., Nugent, C., Mulvenna, M., McClean, S., Scotney, B., Devlin, S.: Evidential fusion of sensor data for activity recognition in smart homes. Pervasive and Mobile Computing 5(3), 236–252 (2009)
6. Intille, S.S., Larson, K., Tapia, E.M., Beaudin, J.S., Kaushik, P., Nawyn, J., Rockinson, R.: Using a Live-In Laboratory for Ubiquitous Computing Research. In: Fishkin, K.P., Schiele, B., Nixon, P., Quigley, A. (eds.) PERVASIVE 2006. LNCS, vol. 3968, pp. 349–365. Springer, Heidelberg (2006)
7. Moeslund, T., Hilton, A., Kruger, V.: A survey of advances in vision-based human motion capture and analysis. Computer Vision and Image Understanding 104(2-3), 90–126 (2006)
8. Nugent, C.D., Finlay, D.D., Davies, R.J., Wang, H.Y., Zheng, H., Hallberg, J., Synnes, K., Mulvenna, M.D.: homeML – An Open Standard for the Exchange of Data Within Smart Environments. In: Okadome, T., Yamazaki, T., Makhtari, M. (eds.) ICOST 2007. LNCS, vol. 4541, pp. 121–129. Springer, Heidelberg (2007)
9. Poppe, R.: A survey on vision-based human action recognition. Image and Vision Computing 28(6), 976–990 (2010)

10. Schuldt, C., Laptev, I., Caputo, B.: Recognizing human actions: A local SVM approach. In: Procs. of the 17th Int. Conf. on Pattern Recognition, ICPR 2004, vol. 3, pp. 32–36. IEEE (2004)
11. Singh, S., Velastin, S., Ragheb, H.: MuHAVi: A multicamera human action video dataset for the evaluation of action recognition methods. In: 2010 7th IEEE Int. Conf. on Advanced Video and Signal Based Surveillance (AVSS), pp. 48–55. IEEE (2010)
12. Tenorth, M., Bandouch, J., Beetz, M.: The TUM kitchen data set of everyday manipulation activities for motion tracking and action recognition. In: 2009 IEEE 12th Int. Conf. on Computer Vision Workshops (ICCV Workshops), pp. 1089–1096. IEEE (2009)
13. Weinland, D., Ronfard, R., Boyer, E.: Free viewpoint action recognition using motion history volumes. Computer Vision and Image Understanding 104(2-3), 249–257 (2006)
14. Wolf, P., Schmidt, A., Otte, J., Klein, M., Rollwage, S., König-Ries, B., Dettborn, T., Gabdulkhakova, A.: OpenAAL – the open source middleware for ambient-assisted living (AAL). In: AALIANCE Conf., Malaga, Spain, March 11-12 (2010)

Harmonizing Virtual Forms into Traditional Artifacts to Increase Their Values

Mizuki Sakamoto, Tatsuo Nakajima, Tetsuo Yamabe, and Todorka Alexandrova

Abstract. Computing technologies allow us to enhance our daily artifacts by adding virtual forms to the artifacts. The virtual forms present dynamically generated visual images containing information that influences a user's behavior and thinking and are usually realized by adding a display that shows visual expressions or projecting some information on the existing artifact. We have designed Augmented Trading Card Game, which adds virtual characters and special effects on the trading cards of the Nintento DS game in order to encourage and provoke more social play of the game.

In this paper, after presenting an overview of the case study that enhances traditional artifacts with virtual forms, we present six values that play an important role in the design of the enhanced artifact.

1 Introduction

Recently, our daily digital artifacts are becoming more and more usual and widely sold commodities. For example, recently televisions developed in Japan have become cheaper and cheaper despite of their excellent product quality and functionality. Also, Android mobile phones are becoming popular recently and a wide range of models and functionalities is offered on the market. However, it is very difficult for the users to distinguish the differences in the phones and make a choice. The fact that the product quality does not become the value for many of us to buy the product shows that we need to consider another way to design daily digital artifacts. However, we found that new furniture and fashion goods attract us every year and they do not become commodities that are sold at cheaper prices with the time. The reason for this is the fact that they offer additional values to

Mizuki Sakamoto · Tatsuo Nakajima · Tetsuo Yamabe · Todorka Alexandrova
Department of Computer Science and Engineering, Waseda University
e-mail: {mizuki,tatsuo,yamabe,toty}@dcl.info.waseda.ac.jp

P. Novais et al. (Eds.): Ambient Intelligence - Software and Applications, AISC 153, pp. 207–211.
springerlink.com

users. Especially, the prices for such products are kept high if the products offer the sense of rarity. Digital technologies are effective to make digital artifacts a usual commodity and as a consequence to make their prices cheaper, but these technologies are also effective to add more values to the products by customizing them for each user. The customization may offer the artifacts more attractiveness, which might lead to the increase in their prices.

Virtual forms are realized by adding displays and by projecting information on the artifacts, and can be changed dynamically according to the current surrounding situation. This approach is promising to enhance daily artifacts, and to offer more values on the artifacts. We have designed Augmented Trading Card Game (Augmented TCG), which adds virtual characters and special effects on the trading cards in the Nintento DS game to encourage more social play of the game.

From the experiences with the design of the case study, we found six values to consider how to offer additional values in the enhanced artifacts with visual forms. The values can be used in the following steps. The first step is to identify the values in the traditional artifacts. Then, they can be used for discussing which values should be added or changed in the enhanced artifacts in order to increase their values. Finally, we consider what kinds of virtual forms can be suitable for making the artifacts richer and more enjoyable.

2 Augmented Trading Card Game

A trading card game is also commonly referred to as a collectible card game, a customizable card game, or CCG. For our purposes here, we will use trading card game (TCG) to refer to all the three varieties of games. In a nutshell, a TCG combines the collectability of trading cards with strategic game play. Typically a player purchases a starter set, containing a playable deck of cards and a manual that includes an explanation of the rules and the mechanics of the game in an introductory fashion. One of the biggest problems faced by any new TCG player is the need for an opponent to truly engage in the game play, as it is extremely unusual for any TCG to feature a solitaire mode. Players usually begin playing with a friend, at a particular location such as a hobby game store that offers organized gaming opportunities and includes a tutorial component, or via an online portal.

Computer-based TCG is also becoming popular, and in our project we make a comparison between the real TCG, and its virtual one running on Nintendo DS. An important conclusion resulting from that comparison is that the computer-based TCG loses a lot of realities offered by the real TCG. For example, the sense of real cards is essential for many TCG players since making and completing collections of cards is a significant fount of pleasure for them. The computer-based TCG also implies some communication limitations, because it allows a player neither to have an eye-to-eye contact, nor to look at or chat with the opponent player.

As described above, although most of the current computer-based TCGs lose the realities of the real TCG, we claim that ubiquitous computing technologies may help to recover these lost realities and encourage and attract players to enjoy the computer-based TCG in a very similar way to the real TCG. Moreover, adding special effects and virtual forms to the computer-based game might increase the excitement of the game even more than the real one.

Figure 1 shows Augmented Trading Card Game that is currently developed in our project. The system extends the trading card game running on Nintendo DS, where two players are usually located in different places while playing the game. In Augmented TCG, the opponent player is represented as a virtual character that is visualized using a tool called MikuMikuDance. The movement of the character is synchronized with the movement of the real opponent player by using MS Kinect, and the behavior of the character is determined by the information retrieved from a biosensor attached to the opponent player, i.e. the virtual character's behavior and emotions reflect the real player's behavior to some extent. In Augmented TCG, two virtual forms are used. The first form is superimposed onto the playing table to show the virtual trading cards and some special effects during the play. The second virtual form is installed on the wall to show a virtual character.

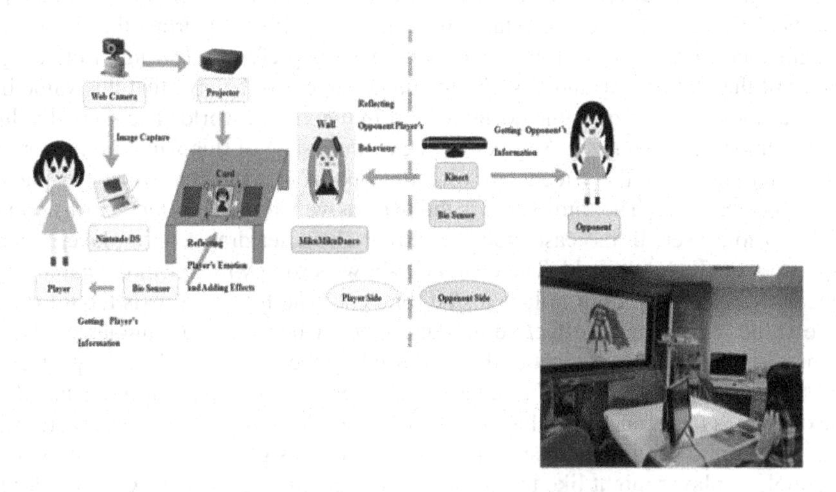

Fig. 1 Augmented Trading Card Game

The trading card itself is also enhanced in the system. Cards presented on the display of the Nintendo DS are retrieved by Web cameras and projected on a real table. The projected cards can be enhanced by adding special battle effects or empathetic effects to the characters shown on the cards.

In the original computer-based game, a player usually cannot see the opponent player. The proposed Augmented TCG enables us to recover this lost reality by adding a virtual character whose movement and behavior are synchronized with the movement and behavior of the real opponent. In addition, the virtual trading cards carry some special effects that increase the sense and the excitement of the battle. Similarly, if the character drawn on a trading card shows some empathic expressions, a player feels empathy with the character on the card, and feels more enthusiastic and committed to the game. These special effects compensate the lost realities of the real trading cards. Also, virtually attached rarity to the virtual trading cards brings a feeling of reality and encourages a strong will to collect virtual trading cards.

3 Design Implications

Baudrillard proposed that the consumption becomes more symbolized and additional values become more important than the products as materials [1]. For example, a brand offers significant additional values to fashion items, and consumers feel the value on their virtual properties. On the other hand, adding values to virtual items makes a user feel the items materialized. This means that real products are becoming more virtual, and virtual products are becoming more real.

From the experiences with Augmented TCG, we extract six values to be used in the design of virtual forms that augment traditional artifacts. The first value is the physical value that offers the tangibility to the artifacts. During the design of Augmented TCG we have found that many players prefer the feeling of the tangibility of the real trading cards while playing a game. We believe that this value increases the reality when some artifacts exist in the virtual world. The second value is the empathetic value. In Augmented TCG, the usage of this value in the virtual character increases the friendship with the opponent character when the player likes the character. The third value is the persuasive value that offers extrinsic motivation to a user. In the case study, a virtual character drawn on a player's card appeares on the battle field. The character shows strong will to win the game, and encourages the player not to lose his/her bravery. The fourth value is the informative value. The value is effective to make a better decision. In Augmented TCG, some detailed information about the card used by the opponent player is projected on the battlefield to provoke a better decision and strategy play in the game. The next value is the economic value. The value is not directly used in the case studies, but we discuss the importance of the value when designing the case studies. For example, a player might like to buy special effects in Augmented TCG in order to improve the play or increase the excitement of the game. Finally, the last value is the ideological value. The value represents the metaphor that shows the dream or the expectation of a user. In the case study, a virtual character, who is a famous and strong trading card game`s hero in an animation story, appears during the game. This encourages the player to imitate the hero and thus become a strong and noble player like him.

The described values are useful to identify what the main values of the traditional artifacts are and how to add additional values to the artifacts for making them richer and more enjoyable. For example, in Augmented TCG, we found that the original game running on Nintendo DS has some problems. For example, since the player does not see the opponent player, he/she tends to easily cheat in the game. We believe that using the empathetic value to the virtual character of the opponent player will prevent from cheating in the game. Moreover, adding special effects to the trading cards are useful to motivate the players to win the game fairly.

During the discussion of the case studies' development, we consider the importance of the economic value. We consider that incorporating virtual items to be exchanged among users is a promising way to motivate users to use the enhanced artifact [2]. For example, if a user develops a new way of customization of an artifact, the other users might be interested to use the customization even if they need to pay some money for it. We believe that this kind of customization may offer an attractive business model to artifacts.

Our approach to add virtual forms to existing artifacts makes it possible to gamify the use of the artifacts and make it more enjoyable. Gamification [3] recently becomes a popular way to make daily and business activities more enjoyable. We hope that the proposed values are also useful to gamify these human activities.

References

1. Baudrillard, J.: Simulacra and Simulation. Univ. of Michigan Pr. (1994)
2. Lehdonvirta, V.: Virtual Item Sales as a Revenue Model: Identifying Attributes that Drive Purchase Decisions. Electronic Commerce Research 9(1) (2009)
3. Zicbermann, G., Cunningham, C.: Gamification by Design. O'Reilly (2011)

Personalized e-Health for Elderly Self-care and Empowerment

Viveca Jiménez-Mixco, Maria Fernanda Cabrera-Umpiérrez,
Alberto Esteban Blanco, Maria Teresa Arredondo Waldmeyer,
Daniel Tantinger, and Silvio Bonfiglio

Abstract. This paper describes the OASIS Health monitoring system, a personalized e-health solution specifically designed for the elderly population in the context of OASIS EU funded project, with the objective to empower the elderly so that they meet their social, emotional and psychological needs, take care for their long-term condition, and prevent further illnesses or accidents. The system integrates four main areas that cover the main aspects of interest related to the health management of the users at their own homes: electronic health record personalization, remote health monitoring, education & coaching, and alerting & assistance.

Keywords: personalization, electronic health record, self-care, empowerment, e-Health, elderly.

1 Introduction

The role of self-care in the management of long term conditions is crucial for effective high quality health care of patients. More and more, patients with chronic diseases and

---

Viveca Jiménez-Mixco · Maria Fernanda Cabrera-Umpiérrez · Alberto Esteban Blanco · Maria Teresa Arredondo Waldmeyer
Life and Supporting Technologies, Universidad Politécnica de Madrid, Avda Complutense 30, Ciudad Universitaria. 28040 – Madrid, Spain
e-mail: {vjimenez,chiqui,aesteban,mta}@lst.tfo.upm.es

Daniel Tantinger
Fraunhofer Institute Fraunhofer-Institut for Integrated Circuits IIS, Am Wolfsmantel 33, 91058 Erlangen, Germany
e-mail: Daniel.Tantinger@iis.fraunhofer.de

Silvio Bonfiglio
FIMI s.r.l, via S. Banfi 1. 21047 SARONNO, Italy
e-mail: Silvio.Bonfiglio@barco.com

P. Novais et al. (Eds.): Ambient Intelligence - Software and Applications, AISC 153, pp. 213–216.
springerlink.com

elderly people that need continuous health surveillance are able to live in their own home and surroundings with help from relatives and health professionals. In order to enable patients to receive a preventative home based self empowered care, a complete system must be provided; it must help them to manage and monitor their daily health status, and needs to be connected to the professional medical system at the hospital. This concept that involves the actions taken towards the possibility of not being continuously attached to the hospital environment is often called "patient empowerment" [1]. It implies a re-distribution of power between patients and physicians and therefore an increase of the individual' autonomy to make informed decisions and personally handle their condition for their own health and well being. This paper presents the approach adopted in Oasis Health Monitoring System (OHMS), which has been developed within OASIS European project [2]. OASIS explored the potential of ICT in all aspects of daily life of the elderly to create ambient intelligence environments that empower them so that they meet their social, emotional and psychological needs, take care for their long-term conditions, and prevent further illnesses or accidents. The OHMS represents an effort to advance in a new generation of telecare services for the elderly devoted to –going beyond the monitoring of the health status- providing them with an easy-to-use system to personally manage their health status and assuring their well-being by means of a constant remote control by professional caregivers.

2 Methodology

The User-centred-design concept has been applied throughout the complete design and development cycle through the OPAF (OASIS Participatory Analysis Framework) methodology [3], a participatory design process specifically created within OASIS framework. Developers were driven by the outcomes of these extensive activities including interviews, surveys and discussions about use cases in various user forums, usability tests, and workshops carried out to involve all the stakeholders (i.e. older adults, informal & formal caregivers, and third parties such as the health care and emergency support service providers). The iterative consensus building among key stakeholders and user groups' representatives enabled the clustering of the functionality of the OHMS in four main areas that cover the main aspects of interest in the area: electronic health record personalization, remote health monitoring, education & coaching, and alerting & assistance. Besides the work performed with users, the main efforts of the development process were focused on customization. Not only must the system fulfil the needs of the elderly in general, but of each specific user, mainly when the subject of the application is the health status. For that purpose, the definition and management of the user's electronic health record (EHR) constituted the central element of the application. The solution for managing the user's health profile implements a distributed system of XML files [4], focusing on personalization, data security and synchronization, and thus ensuring that the stored information is always valid and up-to-date in both sides of the communication channel (elderly and professional clients). The EHR contains critical information related to different aspects of the user's health and implements a pre-defined protocol to automate (transparently to the user) the processes needed for the management of the user's health status, thus adding intelligence to the system, e.g. setting the vital and activity signs that should be regularly controlled, frequency of measurements, or ranges of parameters that define the urgency of response in case of emergency.

3 Results

The proposed solution has been deployed in a distributed way with a client-server approach using Java within Eclipse and Netbeans frameworks, OSGI architecture and Web-service technologies. On the client side, the elderly user application's UI has been carefully designed following the guidelines and the adaptation framework [5] specifically defined in the context of OASIS that take into account different conditions of users such as age, mobility, vision or computer literacy. The elderly are provided with a tablet PC and a set of Bluetooth sensors (3 commercial EC-marked and 3 research prototypes) that will be used to record details of the user's biomedical parameters (SpO2, Heart rate, glucose, etc.) and daily activities (Inactive, Walking, Running, Falling, etc.). All measurements are stored in the user's health record, so that any abnormal situation is detected and sent to the alerting module. The Health Coach module is intended to engage and motivate the elderly with regard to the management of their health status by providing educational content to make them aware of the benefits of a healthier lifestyle and how important it is the adherence to the therapeutic plan prescribed by the doctor to prevent a degeneration of the health conditions. Finally, the system alerts and notifies about an important event, which can be either the receipt of a new message from the doctor, a change in the medication treatment, or, the most important, the warning about a potential dangerous situation. All the information is managed and controlled remotely by the medical doctor on the server side using the Medical Center application, which completes the loop of the concept "e-health for empowerment". It enables the medical doctor not only to perform a continuous tracking of the user's health status, but to be proactive and make decisions on the performance of the user's application by setting the medication treatment and monitoring, defining the educational content and giving them recommendations just as if the user was present at the hospital for a regular visit. The main innovation of the OHMS compared to other telecare systems is that, being integrated in OASIS: a) on one hand the system is able to exchange useful information of other domains (e.g. nutrition, environment, brain skills, transport, etc.) with any other application of the platform through the user's profile, giving the health professional enriched data about the context of the user, and b) on the other hand, external healthcare providers will be able to integrate their services in the platform and connect them to the OHMS with a very small effort. The application prototype has been preliminary tested with 10 Spanish users aged between 55 and 83 in order to get feedback from them related to the usability of the application. The tests, performed by LST-UPM in Madrid, were based on the "Think aloud" and the "System usability scale" techniques [6, 7]. Users were given some time to "play" with the prototype and explore its functionality. By verbalizing his thoughts, the test user enables the developer to understand how he views the computer system. After this time, they filled a System Usability Scale questionnaire related to the complexity, easiness of use and consistence of the system. The analysis of the results gave an acceptable usability score with an average user satisfaction index of 73.5 (on a 1-100 scale). With focus on statements and comments of the participants, the results are being taken into account for further improvements of the application.

4 Conclusions

The European Union is investing thoroughly for the research and development of new healthcare technologies to help empower the patients and specifically the older adults so that they become able to personally take care of their health status and prevent further illness or accidents. This paper presents the work carried out to provide a solution for making the self-care and patient empowerment an effective tool for the management of long term conditions. A personalized e-Health system for empowerment has been developed carefully customized for elderly users, including four main areas that enable comprehensive health management: health profile definition & personalization, health remote monitoring, health coach and alerting & assisting. The system is currently being fully tested in OASIS pilots in several countries in Europe (Italy, Romania, Bulgaria, Germany, United Kingdom and Greece) with hundreds of users, and the results will be extensively analyzed for further improvement of the system.

Acknowledgments. We would like to thank the whole OASIS Project Consortium This work was partially funded by EU in OASIS project (FP7, ICT-2007-215754).

References

1. Bos, L., et al.: Patient 2.0 Empowerment. In: Arabnia, H.R., Marsh, A. (eds.) International Conference on Semantic Web & Web Services SWWS 2008, pp. 164–167 (2008)
2. Bekiaris, E., Bonfiglio, S.: The OASIS Concept. In: Stephanidis, C. (ed.) UAHCI 2009, Part I, HCII 2009. LNCS, vol. 5614, pp. 202–209. Springer, Heidelberg (2009) ISBN: 978-3-642-02706-2
3. Lindsay, S., et al.: OPAF: OASIS Participatory Analysis Framework. In: OASIS 1st International Conference, Florence (November 2009)
4. Esteban, A., et al.: Distributed and synchronized users' profile management for Ambient Assisted Living. In: III WTHS, Valencia (December 2011)
5. Melcher, V., et al.: OASIS HMI: Design for Elderly – A Challenge. In: Rice, V. (ed.) Advances in Understanding Human Performance, Print ISBN: 978-1-4398-3501-2
6. Gould, J.D., Lewis, C.: Designing for usability: key principles and what designers think. Communications of the ACM 28, 300–311 (1985)
7. Brook, J.: System Usability Scale SUS - A quick and dirty usability scale (1996), `http://www.usabilitynet.org/trump/documents/Suschapt.doc`
8. Jiménez-Mixco, V., et al.: The OASIS Health monitoring system. In: OASIS 1st International Conference, Florence (November 2009)
9. Monteagudo Peña, J.L., Moreno Gil, O.: e-Health for Patient empowerment in Europe. Informes, Estudios e investigación 2007. Ministerio de Sanidad y Consumo, Instituto de Salud Carlos III (2007)
10. Clemensen, P.J., Rasmussen, J.: Empowerment and New Citizen Roles through Telehealth Technologies. In: eTELEMED 2011: The Third International Conference on eHealth, Telemedicine and Social Medicine (2011)

Part III
ARCD Session

An Evaluation Method for Context–Aware Systems in U-Health

Nayat Sanchez-Pi, Javier Carbó, and Jose Manuel Molina

Abstract. Evaluations for context-aware systems can not be conducted in the same manner evaluation is understood for other software systems where the concept of large corpus data, the establishment of ground truth and the metrics of precision and recall are used. Evaluation for changeable systems like context-aware and specially developed for AmI environments needs to be conducted to assess the impact and awareness of the users. E-Health represents a challenging domain where users(patients, patients' relatives and healthcare professionals) are very sensitive to systems' response. If system failure occurs it can conducts to a bad diagnosis or medication, or treatment. So a user-centred evaluation system is need to provide the system with users' feedback. In this paper, we present an evaluation method for context aware systems in AmI environments and specially to u-Heatlh domain.

1 Introduction

AmI environments are integrated by several autonomous computational devices of modern life ranging from consumer electronics to mobile phones. AmI has several spheres of application like: Transportation, Health, Education, Business, etc, but recently the interest in Ambient Intelligence Environments has grown considerably due to new challenges posed by society, demanding highly innovative services such as vehicular ad hoc networks (VANET), Ambient Assisted Living (AAL), e-Health, Internet of Things and Home Automation among others. These society challenges force developers to take into account growing demands of users.

Furthermore, the increase in ageing of European population and the treatment of chronic and disabled patients implies a high cost in terms of time and effort. Sometimes patients and also healthcare workers consider treatments in health

Nayat Sanchez-Pi · Javier Carbó · Jose Manuel Molina
Group of Applied Artificial Intelligence (GIAA),
Computer Science Department, Carlos III University of Madrid
e-mail: {nayat.sanchez, javier.carbo, josemanuel.molina}@uc3m.es

P. Novais et al. (Eds.): Ambient Intelligence - Software and Applications, AISC 153, pp. 219–226.
springerlink.com

centres unnecessary as they could collapse national health services and increase costs. On the other hand, we face the problem of the patients living in rural areas, where is difficult to access. To face these challenges we need to differentiate medical assistance in health centres from assistance in a ubiquitous way that it is possible due to the advances in communication technologies.

Systems developed for e-Health environments need to be autonomous and self-managed. They need to adapt not only to changes in the environment, but also to the user requirements and needs. User has to take a relevant role providing an evaluation of the system behaviour while using it or at least once it has been used. One of our goals is to evaluate enhanced user experience in the course of using our system and provide automatic adaptation taking into account changes in user preferences and environment. So, generic user-centred evaluation system provides users with the possibility of having a proactive role when using the system. It is on users hands to provide the system with a feedback of the correctness of the provided services. Users will then be capable of specifying the right or wrong context information at a high-level concept so that the system could learn from it and self-adapt its behaviour for future times.

The rest of the paper is structured as followed. First section we present a general issues on evaluation and metrics for context-aware systems. Later our evaluation proposal is laid down. At the end an u-Health case study is presented and conclusions are outlined.

2 Related Work

Evaluation is a central piece of software engineering. Evaluations methodologies allow researchers to assess the quality of the findings and to identify advantages and disadvantages of a system. The goal in evaluation of conventional systems is to proof that a system is more efficient. Normally, variables associated with efficiency are the time to complete a given task or the number of errors that have been made while fulfilling the task.

However, in AmI when a system augments an environment enabling a user to do new things the metric is not straight forward anymore. So, it is important before evaluating a context-aware system to figure out what is the evaluation goal. Context-aware services must dynamically adapt to the needs of the user and to the current physical, social and task context in which those needs are formed. Developing an effective context-aware adaptive service therefore requires extensive user-centred design and evaluation as the proposed adaptive functionality for the service needs to evolve.

Since a few years, there is a growing interest in understanding specific evaluation problems that arise from context-aware systems [10, 3]. But context-aware systems are designed to provide users with services but where the main point arises in the potentiality of context. This kind of systems is designed to help users in a certain situation and provided information that is useful for a particular task [9]. Evaluation procedure becomes then a difficult problem to face when dealing with this kind of systems. If we assume that the main purpose of context-aware system is to provide a user with information according to the contextual information, there are several problems the user currently has in accessing this information: *Distribution of the information:* concerning the manner in which the information is requested and disseminated to a user.

In AmI environments, user context and user preferences become essential aspects when deciding, which of the available services are of most interest to the user in a given situation [6]. Ubiquitous healthcare (u-health) is an emerging area of technology that uses a large number of environmental and patient sensors to monitor and improve patients' physical and mental condition. U-Health focuses on e-Health applications that can provide health care to people anywhere at any time using information and communication technologies. Besides, innovative approaches in mobile healthcare (m-Health) have also been developed as a footbridge between e-Health and u-Health.

Several initiatives, such as Mobihealth [11], XMotion [4] and MyHearth [5] have investigated the feasibility and benefits of mobile healthcare services and applications. However, these initiatives do not provide an evaluation system taking into account the contextual information as well as the user's opinion. The main contributions of this paper is to provide with an evaluation method for u-Health systems.

3 Our Evaluation Proposal

For evaluating the performance of various context-aware systems we use a three-facet approach. First, we use a taxonomy of pervasive computing systems based on our survey of proposed and prototyped systems and research projects. Second, we create a set of case scenarios which serves as a checklist of goals and functionalities for system designers to consider during both design and implementation stages. Third, we identify critical parameters for evaluating context-aware systems and a list of parameters allowing to decide what and how to evaluate each case scenario.

Taking all these into account, following we present a methodology for context-aware systems in AmI environments based on [1]. The taxonomy was constructed using a bottom-up approach and includes: architecture, modularity, geographic span, purpose and integration criteria, to define categories and key parameters to measure each one. In AmI environments the evaluation process lays the following steps:

1. **STEP 1**- Definition of the purpose of the evaluation: It is where the set of inputs is defined. This step has tremendous importance since it is here where we represent the user needs, standards and the state-of-the-art technologies. This is the reference later to match against the main objectives and intentions of the evaluation. For instance, in AmI it not the same if we evaluate the impact of the user or the performance of the system. The two approaches are lay down in next sections.
2. **STEP 2**- Design: It concerns with the set up of the evaluation plan according to the previous step. Along with the creation of the appropriate tools taking into account for instance, to perform the evaluation in an automatic, objective and respectable way.

3. **STEP 3**- Execution: Involves the measurement of the previously selected characteristics, the comparison by using the selected criteria and the assessment of the results. This step is the one that provide the feedback to the developers to serve the subsequent iteration of the design process.

Once we have decided the purpose of the evaluation, we have designed a plan of evaluation, we need to decided which metrics involve. If we would like to know the satisfaction rating of the user with a particular activity, we can measure the utility of the information delivered on attractions. So in the evaluation process we will focus on the application purpose criteria: assurance, assistive, return of investment, experience enhancement and exploration. The requirements and emphasis on various performance parameters are heavily dependent on their primary purposes. For instance in the case of return of investment, services serves primarily to increase the efficiency of users or the environments they are in, and potentially streamline the routine tasks and remove inefficiency. More prominent in business environment, like U-Commerce. Key parameters of these systems are *speed* and *efficiency*, especially of system response time and failure rate.

Another example is experience enhancement services that focus on enhancing and enriching user experiences while interacting with pervasive computing systems. They provide additional or enhanced opportunities for learning, entertainment, or sensual experience. Existing experience enhancement systems includes scenarios where personalization take place. So, usability and quality of context (QoC) represent two metrics very important for this kind of domains.

Measuring Usability

Metrics for usability are variables that are measurable in an objective manner. These variables are structured in three groups and we detail which one we use for this kind of scenario:

1. Effectiveness: Variables that allow us to measure the accuracy and completeness with which it achieves the objectives of a specific task.
2. Efficiency: Refers to the effort that a user has to do to get a goal.
3. Satisfaction: Refers to those who have more to do with the emotional or subjective.

Measuring QoC

Quality of Context (QoC) was first defined in [2] as "any information that describes the quality of information that is used as context information ". While different types of contexts will have QoC attributes specific to them, there are certain attributes that will be common to most contexts and they are:

1. Precision measures how accurately the context information describes reality, e.g. location precision.
2. Probability of correctness (poc) measures the probability that a piece of context information is correct.
3. Resolution denotes the granularity of information.

4. Up-to-dateness specifies the age of context information.
5. Refresh rate is related to up-to-dateness, and describes how often it is possible or desired to receive a new measurement.

4 Case Study: Evaluation of U-Health System

In order to provide with a user-centred evaluation, we have developed a self reported data toolkit called My feedback, which will help to capture user's context, user's ratings, intentions and actions. These data can be acquired from an offline system or an online system that is installed in the users' device. Although we developed both, it is particularly useful and more accurate when collected during or shortly after key moments of interest while still fresh in the user's mind and they do not require retrieval or reconstruction data from memory but access to and accurate reporting of information available to conscious awareness. My feedback toolkit runs on SmartPhones, PocketPCs, TabletPCs, and desktop machines running Microsoft Windows. Questionnaires are triggered based on the movements of users. To maximize user response, a numerical ratings is employed because it is much more efficient. However, this efficiency is at a cost of losing qualitative data. Thus, in this study, we also use an open question where users can recommend the correct service he would like to receive at that time and at that place. To clarify the system's functionalities, following there is a scenario where this U-health Information System can be found beneficial:

Scenario — Hospital

Hospital room, where a patient is monitored for health and security reasons. Objects in the environment are furniture, medical equipment, specific elements of the room like a toilet and a window. Users in this environment will be the patient, relatives and carers (e.g., nurses and doctors). Sensors can be movement sensors and wrist band detectors for identifying who is entering or leaving the room and who is approaching specific areas like a window or the toilet. Actuators can be microphones or speakers within the toilet to interact with the patient in an emergency. Contexts of interest can be " the patient has entered the toilet and has not returned after 20 minutes " or " frail patient left the room ". Interaction rules can consider, for example, that "if patient is leaving the room and status indicates that this is not allowed for this particular patient then nurses should be notified".

We use a multi-agent approach to implement the system published in previous work [7]. The behaviour of the system is the following: In the first phase, the Aruba Positioning system discovers the user's position(patient, doctors, nurses, medical assistant) while he enters the Wi-Fi network in the Majadahonda Hospital. Later, sensor agent provides user's positioning information to the user agent. Once user agent knows its location sends it to the facilitator agent as well as the information regarding using a specific kind of service. Following the set of phases, the facilitator agent communicates with provider agent, in this case Administration Agent which provides a turn to the Patient's agent to see a specialist doctor that has been also previously detected. Administration agent asks the patient agent about context information (medical condition, vital signs) to be used during the interaction to provide the personalized service. Once this context information is received by the Context Manager included in the provider agent, it loads the specific context profile characteristics. This information is then consulted to personalize the provided service. Following the evaluation methodology for AmI systems and its three steps presented in the previous chapter, we define the purpose of the evaluation:

STEP 1 — Purpose

In the case u-health domain, it is a distributed system who implements, working to assist users of the hospital domain while they are carrying out their activities in different zones and with different preferences and roles. Again, taking into account the taxonomy, the purposes of evaluation in this case are:

- to measure the **application purpose** in terms of *usability*;
- to measure the **intelligence** in terms of context awareness, that means *quality of context*;

STEP 2 — Design

At the impact level, a possible metric would be the satisfaction rating of the user with a particular activity. Quality process has two distinct facets: technical quality and functional quality. Technical quality refers to the accuracy of medical diagnoses and procedures, and is generally comprehensible to the professional community, but not to patients. Patients essentially perceive functional quality as the manner in which the service is delivered; while healthcare professional can be capable of making a technical quality evaluation. There are several proposals regarding service quality measurement. Regarding this, for e-health environment we consider two groups of users: patients/caregivers/patient's relatives and health professionals. The first group will be able to make an online evaluation (OnE) of the system, for which we have defined some quality of context measurement, and the second one, an offline evaluation (OffE) with other quality of context measurement that evaluates, in this case, the technical quality of the system response.

STEP 3 — Execution

First, as the awareness of the system has been adapted for the e-health environment, the evaluation will be done based not only on the patients' location (as we did in [8], but also on his vital signs: blood pressure (BP); pulse rate (PR); respiration rate (RR) and body temperature (BT). We based on the fact that the system is composed of a set of different sensors connected to a PDA that transmits, in a secure way, all the patient data (location and vital signs) to a central server in the hospital. The authorized doctors can access this medical information from their computers (inside the hospital or even outside) afterwards. For offline evaluation, we explore the

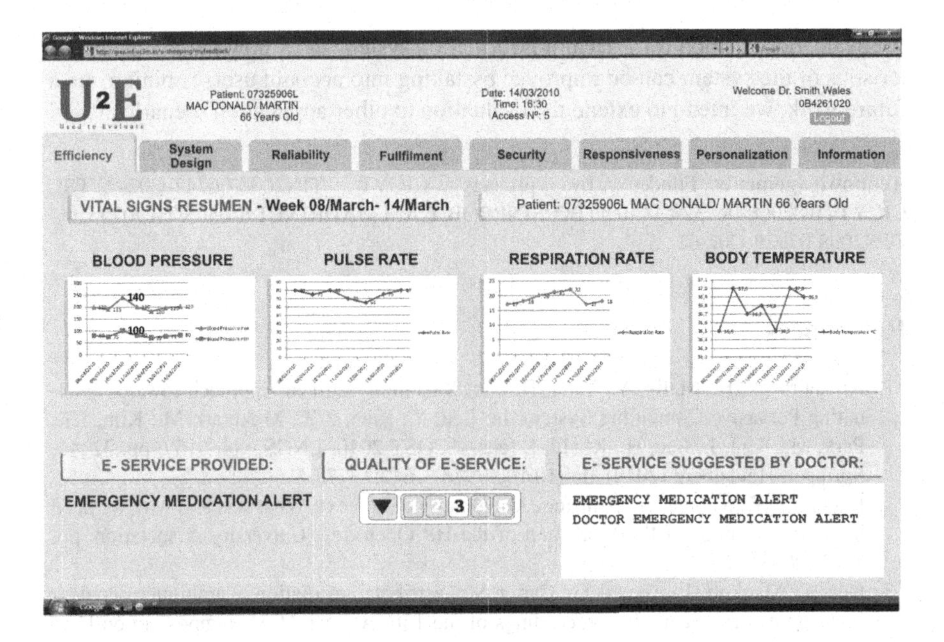

Fig. 1 Myfeedback: Offline evaluation toolkit

e-service quality dimensions based on a review of the development of e-service quality scales. So, once the e-service is provided to the patient, the OnE evaluation system is invoked by the evaluation agent and patient/caregiver/patient relatives can make the evaluation of the e-services received filling the evaluation form. Doctor or health professional in charge of following the patients' file, can also evaluate the system behaviour as see in Figure 1. In this case Dr. makes an offline evaluation of the behaviour of the system during a week. Dr. suggests the system, in a similar case, to activate the DOCTOR EMERGENCY MEDICATION'S ALERT that will send a message to the doctor, so he can be notified immediately.

5 Conclusions

Evaluations for context-aware systems can not be conducted in the same manner evaluation is understood for other software systems where the concept of large corpus data, the establishment of ground truth and the metrics of precision and recall are used. Evaluation for changeable systems like context-aware needs to be conducted to assess the impact of the users. The heterogeneity, dynamicity, and heavily context-dependent behaviors of context aware systems require new approaches of performance evaluation. Normally, apart from the simulation techniques, real-world evaluation is conducted as field studies and relies on collecting data from observation about the usability of the software in the context of use. We have applied an evaluation methodology for this kind of scenarios. The results of the application of the evaluation method for u-Health information system show how the main characteristics of the system can be improved by taking into account users' opinion. As a future work, we intend to extend the evaluation to other application scenarios.

Acknowledgements. Funded by projects CICYT TIN2008-06742-C02-02/TSI, CICYTTEC2008-06732-C02-02/TEC, SINPROB, CAM MADRINET S-0505/TIC/0255 and DPS2008-07029-C02-02.

References

1. Abdualrazak, B., Malik, Y., Yang, H.-I.: A Taxonomy Driven Approach towards Evaluating Pervasive Computing System. In: Lee, Y., Bien, Z.Z., Mokhtari, M., Kim, J.T., Park, M., Kim, J., Lee, H., Khalil, I. (eds.) ICOST 2010. LNCS, vol. 6159, pp. 32–42. Springer, Heidelberg (2010), doi:10.1007/978-3-642-13778-5_5
2. Buchholz, T., Küpper, A., Schiffers, M.: Quality of context: What it is and why we need it. In: Proceedings of the Workshop of the HP OpenView University Association, pp. 1–13 (2003)
3. Dey, A., Mankoff, J., Abowd, G., Carter, S.: Distributed mediation of ambiguous context in aware environments. In: Proceedings of the 15th Annual ACM Symposium on User Interface Software and Technology, pp. 121–130 (2002)
4. Mentrup, C.: X-motion project. Tech. rep., T-SYSTEMS NOVA Gmbh (2004), http://cordis.europa.eu/fetch?
5. Philips Research: Myheart project webpage (2010), http://www.research.philips.com/technologies/heartcycle/myheart-gen.html/
6. Rasch, K., Li, F., Sehic, S., Ayani, R., Dustdar, S.: Context-driven personalized service discovery in pervasive environments. In: World Wide Web, pp. 1–25 (2011)
7. Sánchez-Pi, N., Molina, J.M.: Adaptation of an Evaluation System for e-Health Environments. In: Setchi, R., Jordanov, I., Howlett, R.J., Jain, L.C. (eds.) KES 2010, Part IV. LNCS, vol. 6279, pp. 357–364. Springer, Heidelberg (2010)
8. Sánchez-Pi, N., Molina, J.: A multi-agent approach for the provisioning of e-services in u-commerce environment. Internet Research 20(3) (2010) ISSN: 1066-2243
9. Schmidt, A.: Ubiquitous computing- computing in context. Ph.D. thesis (2002)
10. Scholtz, J.: Ubiquitous computing goes mobile. ACM SIGMOBILE Mobile Computing and Communications Review 5(3), 32–38 (2001)
11. Van Halteren, A., Bults, R., Wac, K., Konstantas, D., Widya, I., Dokovsky, N., Koprinkov, G., Jones, V., Herzog, R.: Mobile patient monitoring: The mobihealth system. The Journal on Information Technology in Healthcare 2(5), 365–373 (2004)

Adaptation and Improvement of the Link, Network and Application Layers for a WSN Based on TinyOS-2.1.1

Antonio Rosa Rodriguez, Francisco J. Fernandez-Luque, and Juan Zapata

Abstract. Due to the increasing interest in collecting data related to physical activity habits and behavior of people, wireless sensor networks have emerged as one of the most appropriate technologies for applications that focus on the interest of health care and assisted living. DIA system (in Spanish, *Dispositivo Inteligente de Alerta*) has developed devices which detect behavior patterns from their users and use them to take alert actions when happen significant variations on these patterns. To sum up, DIA deploys a minimum set of sensors with sensory capabilities via wireless communications interface and according to a predefined semantics. In this paper, we prolong the battery life of wireless devices DIA by means, firstly optimizing the energy consumption of its communications, secondly adapting the firmware that performs the monitoring tasks and processing the information acquired by wireless devices, keeping or improving the quality of service provided. For this, we have used source code libraries that provide a complete source code communication protocol stack and TinyOS-2.1.1 as operative system.

1 Introduction

LR-WPAN (Low-Rate Wireless Personal Area Networks) are part of the WPAN networking group (Wireless Personal Area Network) [1]. Such networks are the only ones that have been designed for typical applications such as wireless sensor networks. In these applications there are two requirements that establish a compromise: network capacity to manage data in real time and enough energetic autonomy in the nodes. With a greater capacity and resources in the processing, transmission and reception of data, more applications will support this technology, although with less

Antonio Rosa Rodriguez · Francisco J. Fernandez Luque · Juan Zapata
School of Telecommunications, Universidad Politécnica de Cartagena
e-mail: {antonio.rosa, ff.luque}@ami2.net,
juan.zapata@upct.es

P. Novais et al. (Eds.): Ambient Intelligence - Software and Applications, AISC 153, pp. 227–234.
springerlink.com

life for the batteries of the nodes. This optimization problem is one that has been taken into account when designing the IEEE 802.15.4 standard [2], which are based LR-WPAN.

The DIA system [3] monitors ubiquitous and unobtrusive activity of old people in order to detect anomalous events or activities [4]. For this, it is necessary to deploy the sensor nodes through the rooms of the home where the user spends most of its time alone. The system uses several sensor nodes for monitoring of the old people activity: motion in the rooms, pressure sensing on beds or sofas and doors openning. All events detected by the sensors are transmitted by the wireless sensor network set up with the technology based on MoteWorks™ [5] platform that implements a protocol stack that allows the creation of end-end wireless sensor networks. The system has a processing unit (PC) that allows continuous monitoring of the user. The processing unit receives events from the sensors via a base station and decides what action is appropriate to take. All this infrastructure is governed by sensory rules provided by an artificial intelligence software embedded in the unit processing that allows a dynamic learning of daily activities of the user. In Figure 1 is shown the sensory system infrastructure.

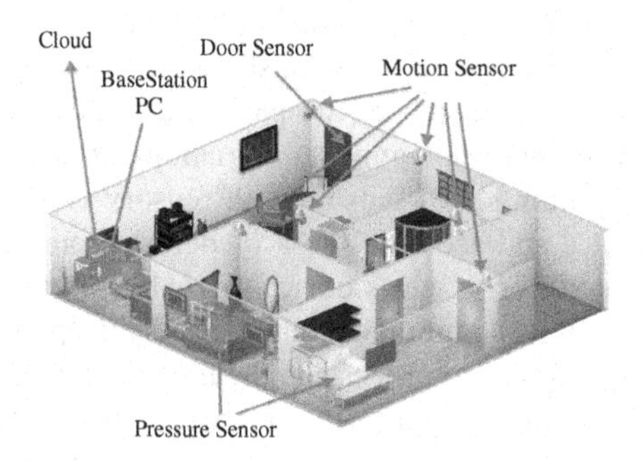

Fig. 1 Schematic view of the system deployed in a house.

This initial version of the DIA system, which deploys a wireless sensor network based on a set of communications protocols (MoteWorks™), has not result as efficient in terms of energy consumption as expected. In addition, these communications protocols are proprietary code, which has been a problem when adapting and enhancement of the DIA system.

The main goal of this work is to increase the battery life of the sensor nodes in DIA's sensor network. It is essential to achieve this goal using communication protocols that optimize energy consumption, keeping or improving the quality of service provided. In Section 2, the wireless network and link protocol employed in the initial version of DIA system is presented and analysed. Moreover, the most important drawbacks of these protocols are commented. In section 3, a new network protocol with a different approach to earlier is presented. After suggesting a new network protocol, a MAC protocol called BoX-MAC is presented. This protocol enhances the MAC protocol used in the initial version of DIA, reducing the consumption of energy and increasing the throughput. In section 4, the phases of the adaptation and implementation process in order to integrate the suggested protocols in the DIA's system are commented. In section 5, results show that our goals are reached. Finally, in section 6 several conclusions and future works are discussed.

2 DIA System: Network Protocol and MAC Protocol

XMesh is a multi-hop network protocol provides by MoteWorks™ to run in the family of MICA and IRIS motes [6] (based on Atmel [7] microprocessors) using the TinyOS [8] programming environment. It is a protocol for ad-hoc mesh networks, with the ability to form networks without human intervention. It also allows adding and removing nodes in the network without having to reset all the network. The base station uses routing beacon messages to establish bidirectional routes between nodes and itself.

The function cost metric of XMesh is based on minimizing the number of transmissions required to deliver a package via multi-hop to its destination, this function is called minimum-cost metric transmission (MT). This metric differs from traditional cost metric distance vector routing based on the number of jumps.

XMesh uses the contention-based protocol B-MAC (Berkeley MAC) [9] as media access control protocol (MAC). B-MAC provides a reduced feature set and an efficient mechanism energy-saving for accessing to the channel: back off, idle channel and estimate channel link. In addition, in order to achieve a small duty cycle uses a scheme of sleep/activity-based asynchronous over a period of low power listening (LPL).

Into the scheme of sleep/activity, once the sensor node is woken up, the sensor node checks the channel looking for activity. The period between two consecutive wake up is called check interval. After waking up one node, the node remains active for a period of time called wake up times, in order to to detect transmissions. While the wake up time is not configurable, the check interval can be defined by the application. B-MAC packets are composed of a preamble and data. The preamble length is at least equal to the check interval duration, so that each node can always detect a transmission input during their wake up time. This approach does not require that the nodes are synchronized. In fact, when a node detects channel activity, it remains active in order to receive the preamble and then data.

After the completion of a successful pilot installation of the DIA system in one hundred different houses for over 2 years, we have analysed the main drawbacks found in the WSN deployed by XMesh. These drawbacks have served as a starting point to implement a new wireless sensor network for the DIA system.

1. XMesh can not be adapted to DIA sensor network efficiently due to XMesh is developed by proprietary code libraries which can not be modified. Besides, it does not allows to select different intervals of work (duty cycle).
2. For the context of DIA, XMesh routing network is not efficient in static indoor environments where the network is formed with few nodes (8 to 10 nodes). In this

case, a star topology is more efficient than a mesh topology. On the other hand, XMesh is sensitive to small variations of energy of the channel because estimates the link quality of a node according to the energy received by neighbour nodes. Therefore, the logical topology of the network experiences many changes due to due to channel variations even though the network topology is static.

3. XMesh uses a proactive approach. XMesh can only update routing paths after a set time period, rather than when node really needs it. By this reason, XMesh generates unnecessary power consumption.

3 Network and MAC Protocol Proposed: TYMO and BoX-MAC

LR-WPAN networks have promoted the investigation of different routing protocols for low-power wireless networks, most algorithms are based on distance vector type (Ad-hoc On-demand Distance Vector, AODV [10]).

The routing protocol DYMO (Dynamic MANET On-Demand) [11] was developed as a simple implementation, substantially extensible, and the result of the combination of the most important components of the algorithm AODV and other reactive protocols. DYMO performs similar to AODV search mechanisms and maintenance of a route using control messages.

TYMO (DYMO for TinyOS) [12] is the implementation of network protocol DYMO in TinyOS, a routing protocol for point to point networks MANETs (Mobile Ad-hoc Networks). It has been adapted to the characteristics of wireless sensor networks, it is available for all hardware platforms, it has no restrictions in the context of communications (as if they do the diffusion or collection protocols), it is stable, allows the inclusion of mobile nodes and is compatible with low power communications policies. TYMO is based on a reactive approach, meaning that it establishes a route to a destination only on demand. Therefore, if the network is static and the links are stable, it does not run any procedure for network control. This control and routing procedure generates high energy consumption in the network because all network nodes are involved in it. TYMO consists of a set of open source libraries, that can be adapted to the peculiarities of any sensor network. In contrast, XMesh code is proprietary and its operation can not be modified. The theory-based routing by hops (TYMO) is more efficient in terms of power consumption in indoor networks where the node density is small, against the routing procedure which uses the link quality estimation for the establishment of routing paths (XMesh).

TinyOS-2.1.1 improves the robustness and efficiency of low-power asynchronous communication by sharing information on the physical layer (detection power in the channel), and link layer (short preambles header IEEE 802.15.4) in radio duty cycle. Therefore, TinyOS-2.1.1 implements a new MAC protocol type that uses information from both layers (cross-layer). The protocol is called BoX-MAC [14] and provides two versions: MAC BOX-I (improvement of B-MAC protocol) and BoX-MAC II (improvement of X-MAC protocol [13]).

BoX-MAC II splits the preamble length in shorter pieces, so that nodes remain less time awake, occupy the channel less time, and to contain the link layer does not affect all neighbours nodes. In turn, confirms receipt of the link-level message with an ACK packet. It allows users to set different periods of sleep/activity according to the destination node. BoX-MAC II mostly uses low-link layer mechanisms, but includes some information from the physical layer: it uses energy detection techniques of the physical layer (CCA, Clear Channel Assessment).

4 Adaptation and Implementation

The main goal of this work is to increase the battery life of the nodes in DIA's sensor network. It is essential to achieve this goal, to use communication protocols that optimize energy consumption, while maintaining or improving the quality of service provided. Under this context, to achieve this goal is necessary the adaptation of proprietary code firmware embedded in the sensor nodes to a new open source version based on the use of open source libraries.

An OSI model (Open System Interconnection) was extracted from firmware embedded in the sensor nodes. In this model, each layer gathers software components that implement services and processes necessary to perform the tasks running on the sensor node (communication, sensoring, signal processing, and so on). The stack was divided into four main layers: physical, MAC layer, network layer and application layer. The transport layer is integrated into the application layer. The physical and MAC layers implement the services required by the network layer. In turn, the network layer provides communication mechanisms for high-level application layer. The application layer access to resources and services provided by lower layers, and implements the processing and transmission of information obtained by the measurement of the node.

Once extracted a layers model, where each layer provides services to adjacent layers, the adaptation process was planned through a approach by stages: in the first stage, all services associated with the application layer was implemented; in the second stage, a new network layer in the application layer was integrated; in the third stage, procedures for medium access and physical transmission of information to the application were implemented; in the last stage, a process of global integration of all layers, debugging and validation was developed.

In the first stage, all services associated with the application layer were implemented again. For it, the application layer services were mapped from TinyOS-1.x to TinyOS-2.1.1. All the state machines, timers, components and other process were mapped to TinyOS-2.1.1 using the NesC [15] language. Moreover, a new format messages was used and new service commands and messages used by all nodes of the network were implemented.

In the second stage, TYMO routing protocol was adapted to requirements of wireless sensor network of DIA system. Due to the wireless sensor network of DIA system implements new low-power procedures in MAC layer (BoX-MAC II), several processes have been reviewed in order to avoid reaching inconsistent states which can block the operation of the network protocol.

In the third stage, MAC level requirements to wireless sensor network were configured. TinyOS-2.1.1 implements the MAC layer protocol called BoX-MAC-II in the communications stack of the transceiver AT86RF230 [16]. High-level components provide basic services protocol applications should run on the nodes. A set of parameters to configure the behavior of the protocol is defined. This set of parameters sets up a period of sleep/activity that results in duty cycles for base station and sensor nodes.

Finally, in the fourth stage, the intelligent artificial system is adapted to TYMO format message and whole system is debugged. The intelligent application, which interprets the information from sensor nodes in order to determine the probability of a dangerous pattern of behavior, is an application programmed in Java, that runs on a PC with OS Linux. The developed firmware for the sensor network must be subjected to a process of debugging and validation. To facilitate this task we use two main tools, one to facilitate real time interaction (SensorReader) and one for the analysis of driver logs (ProdiaMonitor).

5 Results

Each of the planned stages corresponds with the goal of this work and meets all the prerequisites. In short, (1) an adaptation of the link level has been performed, by means of the integration of a MAC protocol (specifically BoX-MAC II) to reduce energy consumption and increase the quality of service provided by the sensor network, (2) an adaptation at the network layer has been performed, by deploying a mesh network of sensors with a new network protocol (TYMO) that meets the DIA's system requirements, and (3) optimizing (in terms of energy consumption and latency) all processes running on the application layer with TinyOS-2.1.1.

To evaluate the increase in the autonomy life of sensor nodes programmed with the new firmware version (TYMO + BoX-MAC II) versus the initial version with XMesh, we have deployed two WSN with identical distribution in a same real scenario during around eighty days. The first one, consists of sensor nodes that are programmed with the XMesh version. The second one, is formed of sensor nodes that are programmed with the new firmware version. All sensor nodes are powered by two AA (1.5 V) batteries type and the minimum voltage value supported by sensor nodes is 2200 mV. The nodes programmed with the XMesh version check the channel with a frequency of 8 times per second, whereas the nodes programmed with the new version check the channel with a frequency of 4 times per second.

The evolution of the voltage level of batteries of both WSN, is shown in Figure 2(a) and Figure 2(b). The horizontal axis shows the elapsed time in days and the vertical axis shows the voltage level in millivolts. In both figures, we have represented the linear regression line adjusts to the evolution of the batteries of the sensor nodes of each network. The TYMO's linear regression function is $y = -4.11x + 2958$ whereas the XMesh's linear regression function is $y = -5.3x + 3051$.

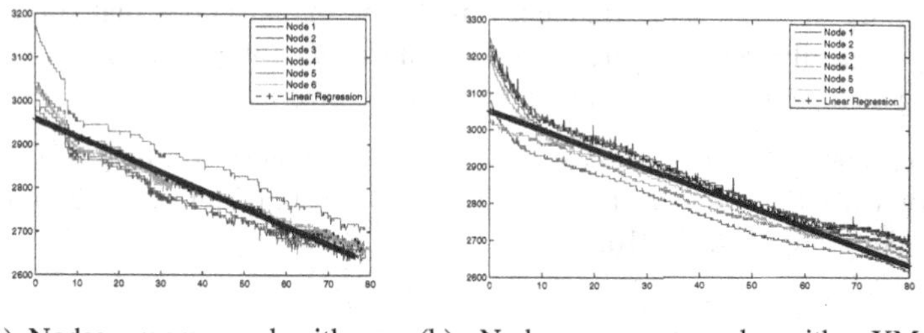

(a) Nodes programmed with new version (TYMO + BoX-MAC II).

(b) Nodes programmed with XMesh version.

Fig. 2 Evolution of the voltage level of batteries and linear regression line estimated for both WSN.

The mean slope of linear regression for each function (in millivolts per day), is a parameter which indicates us the energy average consumption of each network. This parameter shows us that the mean slope of TYMO (-4.11) is below than XMesh (-5.3). Using the mean slope equation of the liner regression line we have estimated the average nodes battery life, and it is beyond the boundary of 3 months achieved by XMesh.

6 Conclusions and Future Works

In this paper, the energy consumption and the quality of service of a wireless sensor network for health care have been improved by means of using open source code libraries and free software components. The quality of service provided has been improved by reducing the delay associated with the establishment of the network and the transmission of messages. With the integration of the TYMO network protocol with the approach of a distributed system, the delays associated with establishing the network and insertion of nodes in a network are minimized. In turn, the delays associated with message transmissions are substantially reduced thanks to improvements in link layer MAC protocol BoX-MAC II. In addition, we have eliminated the problems caused by the use of proprietary code libraries that do not provide interfaces accessible and modifiable. All software resources used in the process of adapting and improving firmware of sensor nodes are based on open source libraries. We have integrated a new network protocol that efficiently adapts to conditions of DIA's sensor network, and adjusts their internal processes to optimize the consumption in the transmission of information and maintaining the network. As well, we have used more efficient transmission schemes, and it has been shown that the MAC protocol BoX-MAC II is more efficient than the protocol used by XMesh (B-MAC).

Upon completion of this work, we propose one improvement that can be the seed for future work. The low-power communications scheme employed, which is based on periods of sleep/activity where all nodes set the same time intervals. However, end nodes (these nodes do not forward traffic from other nodes) do not need listen to the channel with the same frequency as do the parent nodes that support end nodes. Dynamically can configure the periods of sleep/activity nodes in the network, optimizing the overall consumption of the sensor network.

Acknowledgement. This work was supported by the Spanish Ministry of Ciencia e Innovación (MICINN) under grant TIN2009-14372-C0302 and for the Fundación Séneca under grant 15303/PI/10.

References

1. IEEE 802.15.4 WPAN-LR Task Group Website, http://www.ieee802.org/15/pub/TG4.html
2. Zheng, J., Lee, M.J.: A comprehensive performance study of IEEE 802.15.4. Sensor Network Operations, 218–237 (2006)
3. Ambiental Intelligence & Interaction S.L.L. System DIA, http://www.ami2.net/secciones/proyectos.htm
4. Luque, F.J.F.: Wireless Sensor Network System for Assisted Living Home, UPCT (2009)
5. Crossbow Technology. MoteWorks Getting Started Guide, http://www.xbow.com/Support/Support_pdf_files/MoteWorks_Getting_Started_Guide.pdf
6. MENSIC Products. Iris Datasheet, http://www.memsic.com/products/wireless-sensor-networks/wireless-modules.html
7. ATMEL. Atmega1281 Microcontroller, http://www.atmel.com/dyn/products/product_card.asp?part_id=3630
8. TinyOS Documentation Wiki. TinyOS Web, http://www.tinyos.net
9. Polastre, J., Hill, J., Culler, D.: Versatile Low Power Media Access for Wireless Sensor Networks. University of California, Berkeley (2006)
10. Network Working Group. Ad hoc On-Demand Distance Vector Routing. RFC 3561. IETF (2003), http://www.ietf.org/rfc/rfc3561.txt
11. Chakeres, I.D., Perkins, C.E.: Dynamic MANET On-demand (DYMO) Routing draft-ietf-manet-dymo-04. Mobile Ad hoc Networks Working (2006), http://tools.ietf.org/id/draft-ietf-manet-dymo-04.txt
12. Thouvenin, R.: Implementing and Evaluating the Dynamic Manet On-demand Protocol in Wireless Sensor Networks. University of Aarhus Department of Computer Science (2007), http://tymo.sourceforge.net
13. Buettner, M., Yee, G., Anderson, E., Han, R.: X-MAC: A Short Preamble MAC Protocol For Duty-Cycled Wireless Sensor Networks. Department of Computer Science University of Colorado at Boulder (May 2006)
14. Moss, D., Levis, P.: BoX-MACs: Exploiting Physical and Link Layer Boundaries in Low-Power Networking. Rincon Research Corporation, Computer Systems Laboratory Stanford University Stanford, CA (2010)
15. nesC: A Programming Language for Deeply Networked Systems, http://nescc.sourceforge.net/
16. AT86RF230: Low Power 2.4GHz Transceiver for ZigBee, IEEE 802.15.4, 6LoWPAN, RF4CE and ISM Applications (February 2009), http://www.atmel.com/dyn/resources/prod_documents/doc5131.pdf

Topic-Dependent Language Model Switching for Embedded Automatic Speech Recognition

Marcos Santos-Pérez, Eva González-Parada, and José Manuel Cano-García

Abstract. Embedded devices incorporate everyday new applications in different domains due to their increasing computational power. Many of these applications have a voice interface that uses Automatic Speech Recognition (ASR). When the complexity of the language model is high, it is common to use an external server to perform the recognition at the expense of certain limitations (network availability, latency, etc.). This paper focuses on a new proposal to improve the efficiency of the usage of the language model in a recognizer for multiple domains. The idea is based on the selection of a proper language model for each domain within the ASR system.

1 Introduction

In recent years, the interest in the use of voice interfaces on mobile devices has increased. Due to the limited computational power of mobile devices compared to desktop computers, the most common architectures for voice applications (e.g. Google's Voice Search) are NSR (Network Speech Recognition) or DSR (Distributed Speech Recognition) [14]. Thus, NSR and DSR architectures have dependence on the data network and the problem of latency [15]. But due to the rise of smartphones and tablets, the current embedded processors (eg ARM Cortex-A8, Cortex-A9 or Qualcomm Scorpion) achieve performance levels similar to their desktop equivalents of approximately five years ago. So it begins to be possible to perform the complete ASR task on the device.

While several of the internal processes of the recognition task have been heavily optimized to improve performance in embedded systems [10], the complexity of the

Marcos Santos-Pérez · Eva González-Parada · José Manuel Cano-García
Electronic Technology Department, School of Telecommunications Engineering,
University of Malaga, Teatinos Campus, 29071 Malaga, Spain
e-mail: {marcos_sape, gonzalez, jcgarcia}@uma.es

P. Novais et al. (Eds.): Ambient Intelligence - Software and Applications, AISC 153, pp. 235–242.
springerlink.com

language model (LM) is one of the key factors that affect the most the Word Error Rate (WER) and the search time on the recognizer, also called Real Time Factor (xRT).

A possible approach for limiting the LM complexity consists in adapting the LM based on topic classification, which is often called Language Model Switching (LMS). Other authors have used this approach with the aim of improving the WER [11], but they employ various recognizers in parallel or in series. Therefore, although able to reduce the WER, the computational complexity increases. There are other approaches to the adaptation of LM based on the interpolation of LM [3], but they also imply an increase in computational complexity.

We propose an architecture based on LMS, but unlike the other cases, in our approach topic detection and language model switching are performed internally during the recognition process. If the topic classification has a great accuracy, it can be achieved a reduction in both xRT and WER.

This paper is organized as follows: after this introduction, Section 2 describes the proposed architecture of the ASR based on LMS. Section 3 presents the platform were the tests are performed and shows the results of the tests that validate our new approach. Later, in Section 4 we extract some considerations from the results. Finally, in Section 5 we summarize the conclusions drawn from the paper.

2 Architecture Overview

Our approach uses as the baseline system the PocketSphinx recognizer which belongs to the CMU Sphinx family [2]. The reason for this choice is that this recognizer is free and open source, and has been used in real-time applications in embedded systems.

The PocketSphinx recognizer is based on the former Sphinx-II and its internals are described in [10] and [13]. It performs 3 recognition passes:

1. Forward Tree Viterbi beam search: The results are a first hypothesis and a word lattice containing all the words that were recognized during the search.
2. Forward Flat Viterbi beam search: This search is restricted to words identified in the previous step. Like the first pass, the results are a word lattice and a hypothesis.
3. N-best search: It uses the word lattice resulting from previous passes. The result is an N-best list of alternative hypotheses.

The main idea of our proposal is to make the LMS within the 3-pass architecture of PocketSphinx, as shown in Figure 1. During the fisrt pass, a General LM (G-LM) is used to cover all possible topics. Then, topic detection is performed and a Topic Dependent LM (TD-LM) is used for the last two passes.

PocketSphinx employs approximately 70-80% of the time in the first pass and 20-30% in the second and third passes. PocketSphinx also allows to run the first pass of the search frame by frame, so that you can start performing the recognition

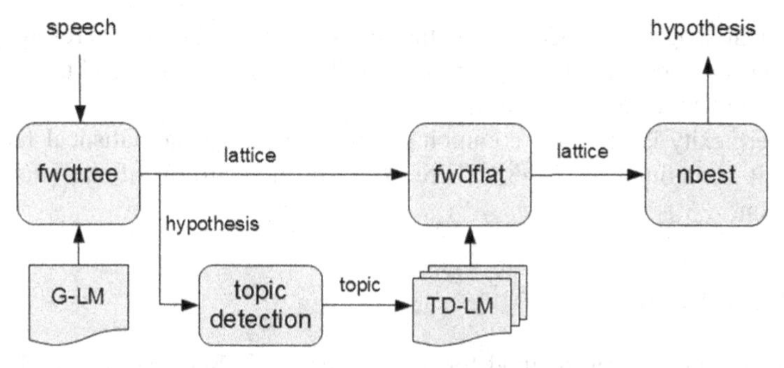

Fig. 1 LMS for embedded ASR. Architecture overview.

while capturing the user voice. This greatly improves the system response time at the expense of worsening the WER, mainly because in this case the cepstral normalization is not performed on the total audio but frame by frame through an estimation process.

2.1 Statistical Language Models

The mathematical model of ASR says that given the acoustic observations O (i.e. the speech features), we should choose the word sequence $\hat{W}$, which maximizes the posterior probability, $P(W/O)$,

$$\hat{W} = argmax_W P(W/O) \tag{1}$$

Bayes rule is typically used to decompose $P(W/O)$ into acoustic and linguistic terms,

$$P(W/O) = \frac{P(O/W)P(W)}{P(O)} \tag{2}$$

and, finally, as $P(O)$ is independent of W,

$$\hat{W} = argmax_W P(O/W)P(W) \tag{3}$$

where $P(O/W)$ is the acoustic likelihood model and $P(W)$ is the language prior model.

Today, n-gram language models are state of the art in applications like speech recognition, machine translation or character recognition. An n-gram is a contiguous sequence of n items from a given sequence of text or speech. In the case of a trigram language model,

$$P(W) = \prod_{i=1}^{n} P(w_i/w_{i-2}, w_{-1}) \tag{4}$$

Statistical language models have the sparse data problem. Therefore, many smoothing methods have been proposed [6]. The best results are obtained using the modified Kneser-Ney algorithm.

The perplexity is the most common goodness measure for statistical language models. It is defined as $2^{H(p)}$, where $H(p)$ is the entropy of the probability distribution.

2.2 Topic Detection

In this paper, the chosen method for topic detection is Support Vector Machines (SVMs) [16]. They have been applied to many different problems and outperform well-established methods such as Artificial Neural Networks and Radial Basis Functions.

SVMs are binary classifiers that try to maximise the margin separation between the data belonging to both classes. For very high dimensional problems in text classification, it is common that the data are not linearly separable in the input space. Thus, SVMs build a decision boundary by mapping the data from the original input space to a higher-dimensional feature space, where the data can be separated using a linear hyperplane.

The selection of hyperplane in feature-space requires the SVM to evaluate scalar inner products in feature-space. But this computationally expensive calculation is not needed due to a functional representation called *kernel*. The *kernel* calculates the inner product in feature-space as a direct operation of the training samples in input-space. Using an effective kernel, this can be done without any significant increase in computational cost.

If the classification problem has more than two classes, it is usual to use the one-against-one or one-against-the rest techniques to transform the multiclass classifier problem in several binary classifiers.

3 Tests and Results

3.1 Test Environment

All tests were executed on a BeagleBoard rev B7 [1], a low power and low cost embedded computer based on Texas Instruments OMAP 3530 system on chip, a variant of the also widely spread ARM Cortex A8 architecture. OMAP microprocessors are quite popular in current embedded computers, since they are the core of many handhelds and mobile phones. Revision B7 of the board is provided with 128 MB of RAM and 256MB of NAND Flash memory, and the microprocessor core is clocked at 600 MHz.

3.2 Test Configuration

We used 3 sets of sentences for testing. Each set comprises utterances belonging to different domains:

- CMU Weather limited domain [8] (*weather* set): This corpus allows weather reports and consists of 100 utterances that cover date, time, outlook, temperature and wind direction.
- CMU Communicator limited domain [7] (*travel* set): This database consists of 530 utterances and is used for the dialog system used in the CMU Darpa Communicator [4]. Communicator is an automated telephone based dialog system for booking flight information.
- DARPA Resource Management Continuous Speech Corpora [12] (*naval* set): The corpus consists of read sentences modeled for a naval resource management task. This database consists of 600 utterances.

Voice recognition needs both an acoustic and a language corpus to run. For this work, we used the Voxforge English corpus [18]. The online available version is a continuous model with 3000 senones. Due to the increased computational complexity that continuous models require [17], it was necessary to perform a new training to obtain a semi-acoustic model with 1000 senones, 256 codebook Gaussians and 5 states per Hidden Markov Model (HMM).

With each of the 3 input sets, *weather*, *travel* and *naval*, we generated a TD-LM based on trigrams using the modified Kneser-Ney method that provides the MIT Language Modeling (MITLM) toolkit [9]. In addition, we trained the G-LM combining the 3 sets ($3topics$). The perplexity of all the language models are shown in the Table 1.

We used for testing the settings' configuration that comes as default for acoustic model training. Table 2 shows the values for the different parameters used in the 3 passes of the decoder.

For topic detection, we represented each utterance S_i that comes from the first pass of the decoder as a point in an n-dimensional space $(O(w_1), O(w_2),, O(w_n))$, where $O(w_k)$ is the number of occurrences of word w_k in S_i and n is the number of total words of the corpus resulting from the union of the 3 domain corpora. For SVM classification we used the LibSVM library [5].

Table 1 Perplexity of the G-LM and each TD-LM.

G-LM (3topics)	TD-LM (weather)	TD-LM (travel)	TD-LM (naval)
4.307	2.699	5.326	3.717

Table 2 Sets of parameters for the decoder.

lw	beam	wbeam	lpbeam	lponlybeam	fwdlw	fwdbeam	fwdwbeam	bestpathlw
10	1e-80	1e-40	1e-40	7e-29	8.5	1e-64	7e-29	9.5

3.3 Results

In these tests, we compared the performance of the baseline recognizer, *LMS_no*, and the recognizer with our proposal of topic detection and language model selection between the passes 1 and 2, *LMS_yes*. For both cases, we measured the average WER and xRT (equations 5 and 6) for all the input sentences, *weather*, *travel*, *naval* and the union of them, $3topics$. In our measurements of RT the time spent in the first pass of the recognizer is deducted because, as mentioned in the Section 2, this part of the recognition takes place while the user is still speaking. Thus, the measured xRT value corresponds to the time spent on recognition once the user completes the sentence.

$$xRT = \frac{Time\ to\ recognize\ the\ input}{Duration\ of\ input\ speech} \tag{5}$$

$$WER = \frac{Substitutions + Deletions + Insertions}{Total\ Words} \tag{6}$$

As seen in in Figure 2, the *LMS_yes* recognizer gets a WER reduction of 15.23% for the *weather* corpus, 9.81% for the case of the *travel* corpus and 1.63% for the *naval* corpus. This results in an improvement in WER for all the union of the corpora, $3topics$, of 3.93%.

Figure 3 shows the results of xRT for both recognizers. Our proposal, *LMS_yes* improves the xRT with respect to the baseline *LMS_no* in 17.51% for the *weather* corpus, 2.37% for the *travel* corpus and 9.76% for the *naval* corpus. Thus, the xRT improvement for the $3topics$ corpus is 8.77%.

On the topic detection, the classifier based on SVMs obtained an accuracy of approximately 99% in the tests.

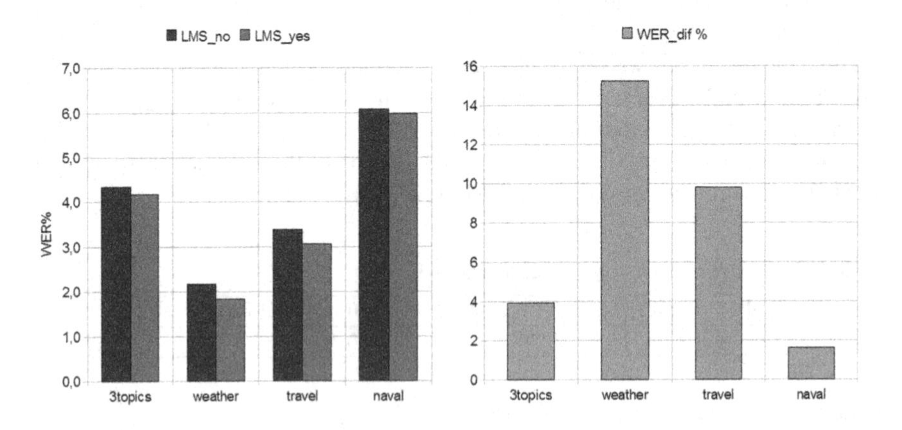

Fig. 2 WER results for each corpus.

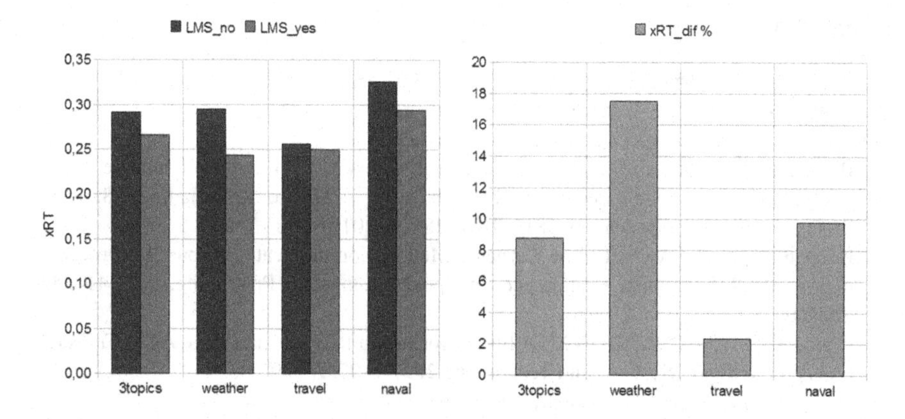

Fig. 3 xRT results for each corpus.

4 Discussion

Based on the results seen in Section 3, we can see that our *LMS_yes* proposal consistently gets better WER and xRT performance than the *LMS_no* baseline recognizer. We can expect that by adding new TD-LMs to the recognizer, the difference in perplexity between the G-LM and the TD-LMs will be wider; therefore, the WER reduction can be much larger. The same applies to the xRT, because the more specific the TD-LMs are with respect to G-LM, the more it gets reduced the complexity of the search in the lattice for the last two passes of the recognizer.

5 Conclusion and Future Work

The main goal of this work was to propose a new approach for improving the performance of multi-domain or multi-task ASR in embedded systems. Thus, we described the motivation of our work and the architecture of our LMS-based recognizer. This was followed by the presentation of promising results that validated our expectations. Finally, we extracted some considerations from the results and established the reasoning that justifies them.

The characteristics of the embedded system used in the tests are similar to those of the vast majority of low and mid-range mobile devices today. For this reason, the results are fully transferable to any mobile device on the market.

In the future, we will address another metrics such as battery consumption and deeper issues related to the parameters of the Viterbi beam search for the different stages of the recognizer.

Acknowledgments. This work was partially supported with public funds by the Spanish National Project TEC2009-13763-C02-01 and by the Andalusian Regional Project P08-TIC4198.

References

1. BeagleBoard website, http://beagleboard.org/
2. CMU sphinx, http://cmusphinx.sourceforge.net/, http://cmusphinx.sourceforge.net/
3. Ballinger, B., Allauzen, C., Gruenstein, A., Schalkwyk, J.: On-demand language model interpolation for mobile speech input. In: Kobayashi, T., Hirose, K., Nakamura, S. (eds.) Proceedings of Interspeech, pp. 1812–1815. ISCA (2010)
4. Bennett, C., Rudnicky, A.I.: The Carnegie Mellon Communicator corpus. In: Proceedings of the International Conference on Spoken Language Processing, pp. 341–344 (2002)
5. Chang, C.C., Lin, C.J.: LIBSVM: A library for support vector machines. ACM Transactions on Intelligent Systems and Technology 2, 27:1–27:27 (2011), http://www.csie.ntu.edu.tw/~cjlin/libsvm
6. Chen, S.F.: An empirical study of smoothing techniques for language modeling. Tech. rep. (1998)
7. CMU Communicator limited domain website, http://festvox.org/dbs/dbs_com.html
8. CMU Weather limited domain website, http://festvox.org/dbs/dbs_weather.html
9. Hsu, B.J., Glass, J.: Iterative language model estimation: Efficient data structure & algorithms. In: Proceedings of Interspeech, pp. 504–511. ISCA (2008)
10. Huggins-daines, D., Kumar, M., Chan, A., Black, A.W., Ravishankar, M., Rudnicky, A.I.: Pocketsphinx: A free, real-time continuous speech recognition system for hand-held devices. In: Proceedings of ICASSP (2006)
11. Lane, I.R., Kawahara, T., Matsui, T., Nakamura, S.: Dialogue speech recognition by combining hierarchical topic classification and language model switching. IEICE - Trans. Inf. Syst. E88-D, 446–454 (2005)
12. Price, P., Fisher, W., Bernstein, J., Pallet, D.: Resource Management RM1 2.0. Linguistic Data Consortium, Philadelphia (1993), LDC93S3B
13. Ravishankar, M.: Efficient algorithms for speech recognition. Ph.D. thesis, School of Computer Science, Carnegie Mellon University, Pittsburgh (1996), Available as tech report CMU-CS-96-143
14. Schalkwyk, J., Beeferman, D., Beaufays, F., Byrne, B., Chelba, C., Cohen, M., Kamvar, M., Strope, B.: "your word is my command": Google search by voice: A case study. In: Neustein, A. (ed.) Advances in Speech Recognition, pp. 61–90. Springer, US (2010)
15. Schmitt, A., Zaykovskiy, D., Minker, W.: Speech recognition for mobile devices. International Journal of Speech Technology 11, 63–72 (2008)
16. Vapnik, V.N.: The nature of statistical learning theory. Springer-Verlag New York, Inc., New York (1995)
17. Vertanen, K.: Baseline WSJ acoustic models for HTK and sphinx: Training recipes and recognition experiments. Technical report, University of Cambridge, Cavendish Laboratory (2006)
18. Voxforge English Acoustic Model website, http://www.voxforge.org/home/downloads

Simulation Based Software Development for Smart Phones

Pablo Campillo-Sanchez and Juan A. Botia

Abstract. Smart phones are getting more and more popular each year and have more and better sensors. The sensors are a rich information source for creating context-aware applications. However, this makes the applications increase in complexity until they become too hard to test in the lab. In order to solve this problem, we propose to test this kind of applications through a simulator. The user interacts with the simulated environment using the keyboard and mouse like a computer game. But the application is not simulated and the user interacts with the application through a smart phone, giving a real experience.

1 Introduction

Mobile applications have experimented a new revolution in the last years. And such revolution has been pushed by two related but, in principle, contrary forces. And they are two different operating systems, iOS and Android. iOS is a closed operating system which is devoted to the mobile devices manufactured by Apple. Thus, the possibilities for developing in such applications are imposed by the Apple policies for open applications development. Android is the operating system designed by Google. And it follows a radically different philosophy of development: open source and Java based development. Moreover, Android is the second most used operating system for smartphones, it is more popular than BlackBerry and iOS, but it is expected that it will overtake to the number one, the Symbian OS by Nokia [3]. So, the mobile software development study is centered on Android mobile applications because the most users are benefited and the group follows an open source

Pablo Campillo-Sanchez
Universidad de Murcia, Campus Espinardo, Murcia, Spain
e-mail: pablocampillo@um.es

Juan A. Botia
Universidad de Murcia, Campus Espinardo, Murcia, Spain
e-mail: juanbot@um.es

P. Novais et al. (Eds.): Ambient Intelligence - Software and Applications, AISC 153, pp. 243–250.
springerlink.com

philosophy like Google. Recently, Smart phones are equipped with a set of sensors, such as GPS sensor, accelerometer senor, gyroscope sensor, camera, microphone and etc. All this hardware allows to get context information about the user is involved into, such as date and time, location, activity. This information is used to develop context-aware applications that offer services to users depending on their needs. Despite the promising potential of using mobile phones as context source devices to make context-aware applications, some problems emerge that need to be solved. One of the central problems on context-aware application development is verification and validation of such applications by testing. Testing software is the process of executing a program in order to find errors in the code [5]. Such errors must then be debugged. According to the IEEE Standard Glossary of Software Engineering Terminology [4]: "Validation is the process of evaluating a system or component during or at the end of the development process to determine whether it satisfies specified requirement". The main challenge of testing is to generate a set of values for each sensor, useful and meaningful for each test. The problem increases with the number of sensors and their correlation. For example, suppose a mobile service that offers indoor location information to others based-location applications like in this work [2]. The service uses the digital compass to get orientation changes and the accelerometer to deduce displacements. To test the service we could generate several sceneries defined by a sequence of values, accelerations for accelerometer and radians for digital compass. Given a initial location and the sensor's values, the service estimates a final location. But it could be easier and more useful if the values are generated indirectly by moving a user and his smart phone in a simulated environment. In this way, the test set could be formed by a list of rooms a user has to visit. Other way could be to define an autonomous user behavior that uses a phone in his natural environment. Anyway, the last two options to define tests are much more natural, comprehensible and realistic than the first one that directly defines a displacement as a sequence of acceleration and radian values. This work is focused on testing of applications or services for smart phones through a simulator where the environment and interactions are modeled. Considering an Android service or application as the system under test (SUT), some of the errors may be found by using a Unit approach [8]. Concretely, UbikSim is used as simulator. It [1] has been designed to simulate environments, domotic devices and people interacting with real ubiquitous software. It is already been focused as an ubiquitous computing environments simulator which tries to alleviate the particularities of testing services and applications whose behaviour depends on both physical environment and users. Moreover, UbikSim offers a world editor called UbikEditor3D. It offers an easy way to create environments and the agents that interact in it. With the simulator not only a sequence of values can be simulated, but it is generated indirectly through defining concrete simulated environments from the reality. It offers two main advantages: (1) software testing is not defined by the component level as a sequence of sensor values, but from stress concrete situations in the virtual world that are more natural and realistic. (2) A graphic representation of the situations means that a set of final users can understand and, therefore, they can to validate the application behaviour more easily while the testing process is performed simultaneously. So, because the user is getting involved in early stages, we achieve a user centering development philosophy in a natural way. In this paper the contents are exposed as follows. Section 2 is related with the challenges. At the next section 3, it is exposed a complex example application to test. Section 4 covers SUT testing by simulation. And, finally, the conclusions and future work are treated.

2 Challenges and Testing Tools for Android Applications

The most common Android development tools [7] are composed by SDK (Software Development Kit) and ADT plug-in (Android Development Tools), both supported by Google and open source. The SDK includes the Android APIs (Application Program Interface), development tools and the Android Virtual Device Manager and Emulator. ADT plugin extends the SDK functionalities to a IDE (Integrated Development Environment). The Android software stack is composed by several layers. The Application Framework Layer is the most important. It provides the classes used to create Android applications and a generic abstraction for hardware access and manages the user interface and application resources. The following subsections contain some Android services that are offered by classes of such layer. They can be used as a source of context for developing context-aware applications. Also, it is explained why the existing testing tools are insufficient to test each application based on such services, and even more if we want to simultaneously test an application based on a correlated set of those services.

2.1 Location-Based SUT

Mainly on a smart phone, location services are obtained through CellID or GPS. The second one is the preferred method as it is more precise but it only works outdoor. CellID approximates your location based on the urban cell you are in, and this could imply too many meters, but it works indoor also. Location services on Android are obtained through the *LocationManager* class. An interesting facility on Android is a hook which allows the programmer to simulate in the emulator, different locations a user passes through, when debugging the application. In this way, the user is not moving, but the emulator virtually does. But the test is not realistic because consists in a file with coordinates and the simulation reproduces constant velocity and straight line displacements. It could be more realistic if the coordinates were given by simulated person displacements in his natural environment. Furthermore, a based-location service including inertial devices could not be tested by this tool.

2.2 Sensor-Based SUT

Sensors on Android are managed in a similar way that location, through a *SensorManager* class which is the one giving access to all the sensors of the phone. Exists a wide list of the sensor-types currently available; note that the hardware on the host device determines which of these sensors are available to the SUT. There are third-party tools which help working with sensors on Android. For example, the Sensor Simulator is a stand-alone application of OpenIntents[1] and it lets simulate moving the mobile and the corresponding sensors by only moving the mouse. Another Sansumg sensor simulator, AndroSensor[2], lets simulate the registers of sensors, obtained by a simulated mobile. It also lets connect to a real device to log real registers from it. But again, those tools are complicated either to manage or to generate the sceneries composed by incompressible sequences of values.

2.3 Audio and Video-Based SUT

Audio and video are another source for context sensing. They are interesting by means of their processing, p. e. image processing to detect objects or recognize commands by speech. Android offers this type of services that support data processing through *Camera* and *AudioRecord* classes for video and image processing, respectively. The emulator allows using a microphone and a webcam to test applications that use those resources. But, it is not considered a tool to feed the SUT with artificial images and sounds that defines scenery for testing.

3 A Hard to Test SUT Example

In this section is studied a type of application which is complex to test. It uses an indoor location-based service (ILBS) to develop an augmented reality (AR). It is conceived to make museum tours more attractive and educational by locating POIs (Point Of Interesting). AR applications need to know the location and orientation of the phone in order to show the POIs on the screen. There are no problems outdoor because the GPS give us the location, but it is not as easy at indoor. In fact, a lot of techniques which try solving this problem are based on different technologies (WiFi, bluetooth, ultrasound, inertial sensors) or a mix. There exist different variants of AR, this paper defines it as a term for a live indirect view (through mobile screen) of a physical real-world environment whose elements are augmented by virtual phone-generated POI icons - creating a mixed reality. The augmentation is in real-time and in semantic context with environmental elements. In order to show the POI icons, each one has a coordinate location and the smart phone gets its own location and orientation from indoor location system. A practical positioning and tracking solution for users in indoor environments relies on both an accelerometer and a digital compass. When a user starts to move, classification data acquired from both sensors are used to approximate the users location. But a mechanism is needed to get an initial position and to solve accumulated sensor errors. So, several QR

[1] http://code.google.com/p/openintents/wiki/SensorSimulator

[2] http://www.newsamsungandroid.com/2011/08/androsensor-all-in-one-android-sensor.html

(Quick Response) codes, with location information, are distributed in the museum. The AR application is developed for the *Museo Arquelogico de Murcia*[3] (MAM). The museum visitors download it and they can browsing through MAM identifying POI, e.g. archeological pieces, next exhibition hall, toilets, etc. So, a user can to identify POIs around him with his smart phone and gets information about them pushing on each POI icon or reaches them physically. Usually, testing stages are divided in several tasks depending on modular functionality of the SUTs. For this example: (1) Friendly graphic user interface (GUI), (2) read QR codes with smart phone camera, (3) the correct QR codes content depending their location, (4) the right location of the POIs and their content, (5)the error of the predictions of the indoor location system based on both an accelerometer and a digital compass and (6) the location of the QR codes depending the error of the predictions. Given these tasks, the fifth is the harder to test. Due to the correlation of sensor values, accelerations and angles. And this type of tests is hard to generate and manage with the actual tools available, as we have seen. To test indoor AR applications it is used a pilot test that is formed a set test in a real environment, a museum in this case. But it is expensive due to it requires to deploy the infrastructure (QR codes, internet access), probably at least an exhibition hall must be closed, time and money to manage and coordinate people. By its cost, a pilot test is usually performed at the end of the development process. At the same time, it implies more costs because the detected errors are more expensive to resolve in this stage than in early ones.

4 SUT Testing by Simulation

Context-aware SUTs are harder to test in a lab as the use of sensor values are more correlated. A simulation-based testing is proposed where environment and its elements related (people and devices included) with the SUT are modeled and simulated. So, first a model of the world and the related elements have to be created in order to the SUT could be tested using a smart phone and the simulator.

4.1 Modeling Elements

In this stage, the simulated world, where the SUT of smart phone will be involved, is modeled. It includes: environment, people and devices. We use for this task an UbikSim tool, UbikEditor3D. Figure 1 shows the editor and a model of an exhibition hall of the MAM. This tool offers an easy way to create environments by dragging elements from the catalog (panel located at top-left) to the panel of edition. It already has some elements but we can to create new ones quickly. In addition, a 3D view of the model is available on the bottom panel. Each world model represents a test configuration. So, first a basic environment is created (e.g., composed by exhibition

[3] http://www.murciaturistica.es/museos/museos.inicio?museo=museo-arqueol%F3gico-de-murcia-(mam)&id=1

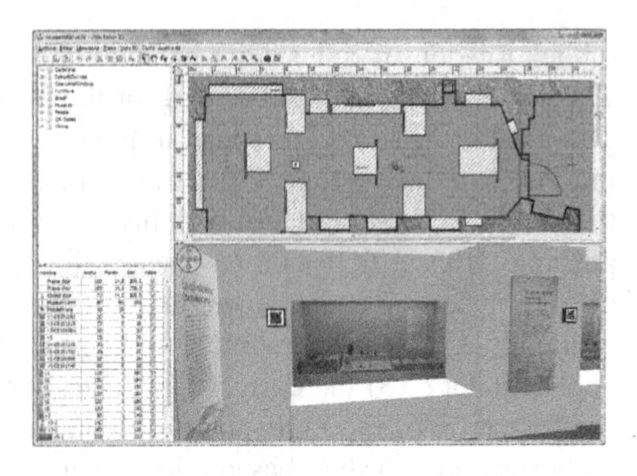

Fig. 1 MAM Exhibition hall modeled with UbikEditor3D

hall, furniture and pieces of art) and then, others more complex to test specific functionalities. For instance, to test the indoor location system, it is shown the predicted tracking on screen to check if the error gotten is acceptable. Other sceneries could be composed by different locations of the QR tags.

4.2 Testing Process

In this stage, a user or developer tests the SUT installed in a smart phone (or emulator) in a simulated world where the user interacts using a keyboard and a mouse. As simulator, we use UbikSim [4] [6] and it has several main features. It has basic models for physical environments (e.g. offices building floors), for humans (e.g. professors in universities) and for sensors (e.g. presence, pressure and open door sensors) that are already developed and validated. Figure 2 shows the main schema of the elements and their interactions needed to test the SUT. At left side, it is the simulator and it includes simulated smart phone (SSP) and simulated user (SU). SU carries SSP and it has simulated sensors that register context information from simulated environment (e.g. simulated temperature sensor registers ambient temperature) and SU actions (e.g. simulated accelerometer registers user displacements). At the right side, we find the real elements. A real user (RU) tests the SUT installed in a real smart phone (RSP) that receives sensor values from the simulated environment through SSP like it comes from the real world. RU interacts with the simulated environment with the keyboard and mouse like a computer game, e.g. Counter Strike[5]. The example exposed in section 3 will be used in order to illustrate how to test an application. It supposed that the simulator represents the predicted location by RSP and the SUT is completely developed. So, the SUT is installed on the RSP. Also, complete scenery is already created and available to be simulated. Once the simulator is started with the scenery and the SUT is installed on the RSP, the test can be begun by the RU. To start, RU could to move his SU until a QR tag using the

[4] UbikSim website: *http://ants.dif.um.es/staff/emilioserra/ubiksim/*. The website offers several videos to view the operation and evolution of UbikSim. It is planned that UbikSim becomes open source. At the moment is a proprietary tool made by Universidad de Murcia and the AmI2 SME.

[5] http://www.counter-strike.net

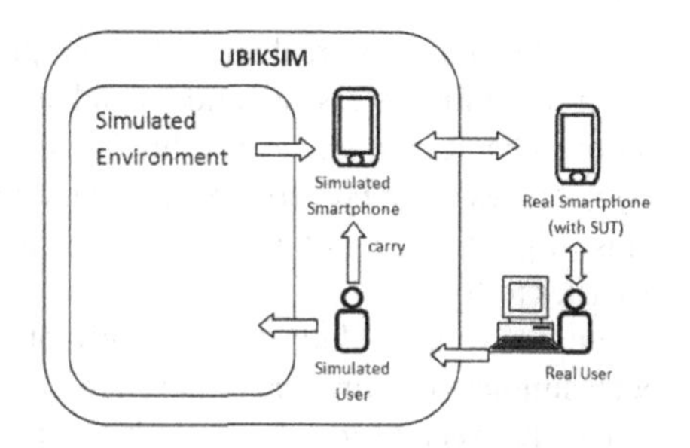

Fig. 2 Proposal of interactions within a mobile applications testing scenario based on simulation.

keyboard, then RU activates the QR reader from his RSP to decode the tag. In order to perform this task, the SUT needs to get images from the camera, instead it receives images (containing the QR tag if it is focused) that are displayed on the PC screen by simulator. The tag is identified and processed to get the location information, after this, it is displayed in the simulator. By this way, the user tests easily if the QR tag contains a correct location. Once the SUT gets its position, the RU can activate the AR from the RSP to identify POIs. RU sees them on RSP screen and can test how the POI icons change by rotating his SU using keyboard. Therefore, RU can to check if POI icons are correctly displayed in our RSP screen and also, RU can to review the content of a POI pushing its icon. Finally, by moving SU and consequently its attached SSP that sends simulated acceleration and orientation changes to the RSP. RSP tries to predict the SSP location from those simulated values and, at the same time, RU checks how varies the predictions on the screen. In addition, RU can check if it affects to AR too much depending on if the POIs are located more or less correctly on the RSP screen. As we have seen, UbikSim contributes to test by means of the displays. The simulation displays are very useful to observe that the application behavior is appropriate. UbikSim works over MASON and can use its features as, for example, inspectors. They are a means to graphically visualize the evolution of variables of interest for the simulation. A large number of inspectors for various simulation variables can be used and monitored dynamically as the simulation evolves. They can be used to check that such variables take always reasonable values, such as estimated locations.

5 Conclusions and Future Work

Currently, existing tools are insufficient to test context-aware applications that make extensive use of their sensor correlated values. This paper proposes a way to test this kind of applications through UbikSim. The user interacts with the simulated environment using the keyboard and mouse like a computer game. But the application is not simulated and the user interacts with the application through a smart phone, giving a real experience. It is a contribution to the engineering process of smart phone application in general and AmI systems in particular. It has three main advantages. (1) The software testing are not defined by the component level as a sequence of sensor values, but from stress concrete situations in the virtual world that are more natural and realistic. (2) A graphic representation of the situations means that a set of final users can understand and, therefore, they can to validate the application behaviour more easily while the testing process is performed simultaneously. (3) Some tests can be performed in early stage of software development process because a pilot test is not needed when before was. When a error is detected early, it is easier to fix.

Future works include a deep study of how to reproduce a test that had already been simulated. It implies to have a mechanism to record a test that includes interactions between user and smart phone. For after, if a test finds an error, it could be reproduced over and over until the error is corrected. In this recorded simulation will be not needed a person and it will be simulated faster, saving time and money.

References

1. Campuzano, F., Garcia-Valverde, T., Garcia-Sola, A., Botia, J.A.: Flexible Simulation of Ubiquitous Computing Environments. In: Novais, P., Preuveneers, D., Corchado, J.M. (eds.) ISAmI 2011. AISC, vol. 92, pp. 189–196. Springer, Heidelberg (2011)
2. Chon, J., Cha, H.: Lifemap: A smartphone-based context provider for location-based services. IEEE Pervasive Computing 10, 58–67 (2011)
3. Gartner Corporation. Gartner says worldwide mobile phone sales grew 35 percent in third quarter 2010; smartphone sales increased 96 perce. (November 2010), http://www.gartner.com/it/page.jsp?id=1466313
4. I. O. Electrical and E. E. (ieee). Ieee 90: Ieee standard glossary of software engineering terminology (1990)
5. Badgett, T., Myers, G.J., Sandler, C., Thomas, T.M.: The Art of Software Testing, 2nd edn. Wiley (2004)
6. Garcia-Valverde, T., Serrano, E., Botia, J.A., Gomez-Skarmeta, A., Cadenas, J.M.: Social simulation to simulate societies of users inmersed in an ambient intelligence environment (2009)
7. Meier, R.: Professional Android 2 Application Development, 1st edn. Wrox Press Ltd., Birmingham (2010)
8. Osherove, R.: The Art of Unit Testing: With Examples in.Net, 1st edn. Manning Publications Co., Greenwich (2009)

Author Index